COMMISSION OF THE EUROPEAN COMMUNITIES

Directorate General for Science, Research and Development
Directorate General for Energy

Energy 2000

A reference projection and alternative outlooks for the European
Community and the world to the year 2000

Jean-Francois Guilmot
David McGlue
Pierre Valette
Christian Waeterloos

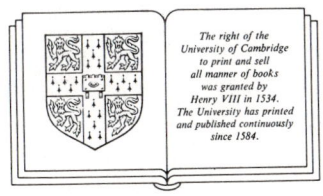

The right of the
University of Cambridge
to print and sell
all manner of books
was granted by
Henry VIII in 1534.
The University has printed
and published continuously
since 1584.

Published on behalf of
the Commission of the European Communities by
CAMBRIDGE UNIVERSITY PRESS
Cambridge
London   New York   New Rochelle
Melbourne   Sydney

Published by the Press Syndicate of the University of Cambridge
The Pitt Building, Trumpington Street, Cambridge CB2 1RP
32 East 57th Street, New York, NY 10022, USA
10 Stamford Road, Oakleigh, Melbourne 3166, Australia

First published 1986

Printed in Great Britain at the University Press, Cambridge

Library of Congress cataloguing in publication data
Energy 2000
1. Power resources.   2. Power Resources–European Economic
Community Countries
I. Guilmot, Jean Francois.
II. Commission for the European Countries.
Directorate General for Science, Research and Development
III. Commission of the European Communities.
Directorate General for Energy

TJ163.E54 1986   333.79   86-17579

British Library cataloguing in publication data

Energy 2000: a reference projection and alternative outlooks for
the European Community and the world to the year 2000
1. Power resources – European Economic Community countries
I. Guilmot, Jean Francois   II. Commission of
the European Communities
333.79'094   TJ163.25.E86

ISBN 0 521 33368 7

Document No. EUR 10367
Commission of the European Countries
Directorate General Telecommunication, Information Industries
and Innovation, Luxembourg

CONTENTS

Preface   vii

Acknowledgements   ix

SUMMARY

I INTRODUCTION   1

II METHOD   2

III A REFERENCE PROJECTION FOR THE EUROPEAN COMMUNITY:
MAIN ASSUMPTIONS   3
(i) General economic framework   3
(ii) Energy context   5

IV A REFERENCE PROJECTION OF FINAL ENERGY DEMAND   6
(i) Background   6
(ii) Overall projections to the year 2000   6
(iii) Sectoral projections   6

V GROSS ENERGY CONSUMPTION TO THE YEAR 2000   10

VI AN ENERGY SUPPLY SCENARIO CORRESPONDING TO THE
REFERENCE PROJECTIONS   12
(a) Historical background   12
(b) Energy production in the Community   12
(c) Energy imports from third countries   13
(d) Electricity production   14

VII MAIN FEATURES OF THE REFERENCE PROJECTION,
COUNTRY BY COUNTRY   16
(a) Belgium   16
(b) Denmark   18
(c) Germany   19
(d) Greece   21
(e) France   22
(f) Ireland   24
(g) Italy   28
(h) Luxembourg   28
(i) Netherlands   29
(j) United Kingdom   31

VIII A STUDY OF SOME ALTERNATIVES TO THE
REFERENCE PROJECTION   32
(a) Variations in the economic framework   33
    (i) economic growth   33
    (ii) economic structure   33

(b) Variations in the average crude oil import price   34
    (i) low oil price   34
    (ii) high oil price   35
(c) Alternatives regarding natural gas prices   36
(d) Alternatives regarding the role of solid fuels   37
    (i) strict emission controls on large industrial plants   37
    (ii) a slowdown in nuclear programmes   38
(e) Analysis of energy efficiency   39

IX THE COMMUNITY'S PLACE IN THE WORLD   40
    Annexes   43

CHAPTER I – REFERENCE PROJECTION: KEY ASSUMPTIONS
A. GENERAL ECONOMIC FRAMEWORK   47
 I Introduction   47
 II The Economy   47
    (i) The international environment   47
    (ii) The European economic outlook   48
    (iii) Sectoral trends   50

B. FUTURE ENERGY PRICES   51
 I General Considerations   51
 II Crude Oil and Oil Products   52
    (a) Cost of imported crude   52
    (b) Consumer prices   55
 III Natural Gas
    (a) Natural gas supply prices   57
    (b) Prices of natural gas delivered to customers   59
 IV Solid Fuels   61
    (a) The cost price of coal   61
    (b) Consumer coal prices   64
 V Electricity   65
    (a) Cost of electrical generation   65
    (b) Consumer prices
    Annexes   69

C. THE COMMUNITY'S ENERGY SUPPLY IN THE LONG TERM   71
 I Energy Supply in 1990   71
 II Energy Supply in 2000   72
    (a) Natural gas supply   73
    (b) The supply of solid fuels   74
    (c) Nuclear energy   76
    (d) Oil   76
    (e) New and renewable energy sources   77

CHAPTER II – REFERENCE PROJECTION: FINAL ENERGY DEMAND
 I The Overall Picture   79
 II Industry   80
 III Transport   82
 IV Residential and Tertiary Sector   83
 V Non-energy Uses   86

CHAPTER III – REFERENCE PROJECTION: ENERGY SUPPLY
 I Trends in Gross Inland Energy Consumption   87
 II The Reference Scenario for Energy Supply   89
    (a) Trends to date   89

(b)   Energy production in the Community   89
(c)   Energy imports from non-Community countries   90

III Main Sectoral Trends   91
(a)   Oil   91
(b)   Natural gas   92
(c)   Solid fuels   93
(d)   Electricity   93

CHAPTER IV – REFERENCE PROJECTION: MAIN FEATURES FOR EACH
MEMBER STATE
Belgium   97
Denmark   110
Germany   122
Greece   134
France   143
Ireland   156
Italy   167
Luxembourg   179
Netherlands   186
United Kingdom   199

CHAPTER V – THE COMMUNITY'S PLACE IN THE WORLD
ENERGY MARKETS
A Introduction   211
B World Energy Consumption: Reference Projection   212
C World Energy Consumption: Reference Projection   217
(a)   Solid fuels   218
(b)   Oil   218
(c)   Natural gas   222
(d)   Nuclear energy   223
(e)   Hydro power and other sources of energy   223

D The Uncertainties on the World Energy Market   224
(a)   Energy demand   224
(b)   Energy supply   225

E Balancing World Energy Demand and Supply in the Year 2000   226
Annexes   229

CHAPTER VI – SOME ALTERNATIVES TO THE REFERENCE PROJECTION
A Introduction   234
B Economic Variants   235
(a)   Changes in the rate of economic growth   235
(b)   Changes in the structure of the economy   237

C Energy Price Variants   240
(a)   Crude oil import prices   240
(b)   Natural gas prices   246

D Alternatives Relating to Demand for Solid Fuels   250

E Different Assumptions About Energy Efficiency   257

CONCLUSIONS   259

PREFACE

The major changes in the world of energy since the first energy crisis in 1973 have impacted heavily on the European Community.

Only 12 years ago oil accounted for 60% of the energy needs of the Community, nuclear energy played only a marginal rôle, coal use was in decline and large amounts of energy were being wasted.

Today oil represents less than 45% of our energy needs and - thanks to the development of the North Sea - imported oil provides less than one-third of the energy we use. Nuclear power has been making major advances in several Member States and in 1985 provided some 30% of our electricity, the equivalent of 12% of our total energy requirements. The use of solid fuels for electricity generation has also increased significantly. Last, but very far from least, energy is being used much more efficiently. We can now produce one unit of GDP with 20% less energy than in 1973.

Few if any of the energy experts at the beginning of the 1970s foresaw such major changes. For most market operators, cheap oil was seen as part of the natural economic order and energy projections tended to be based on simple extrapolation of the increasing use of oil and energy experienced in the 1960s.

The events of 1973, and then the further increase in oil prices in 1979, radically changed perceptions. They upset the comfortable assumptions of the past and forced governments to reexamine their policies and economists to review their analytical tools. The traditional approach no longer provided the key to an energy future which now seemed much less certain.

Recognising the need for a new approach, the Commission of the European Communities launched research work on Energy Systems Analysis and Modelling under the very first Community Non-Nuclear Research and Development Programme in 1975. The Commission's aim was to encourage, facilitate and coordinate research in Member States in an area recognised by everyone as of considerable importance. The Commission also wanted to participate alongside Member States in the development of the most effective analytical techniques.

This work has now borne fruit. It has led to the establishment of compatible energy models for all the members of the Community. And it has made it possible to investigate in a systematic way the links between energy and the rest of the economy, long-term energy demand and the interactions of energy demand and supply.

Energy 2000 – a study of the energy outlook for the Community to 2000 – is the first major product of this investment. It provides detailed and consistent energy projections on a country basis and at the level of the Community as a whole, using throughout the same methodological approach and harmonised energy data and balances. These projections are not of course predictions. Experience has taught us the need for a good deal of humility in analysing the future. They should be considered rather as an attempt to narrow down the range of uncertainties about Europe's energy future. This range depends critically on the future course of economic growth, energy prices, availability of supplies and technological change.

Energy 2000 has already been used as a point of reference for the the reflections of the Commission of the European Communities on energy policy. Its conclusions helped in particular in the definition of the Commission's proposals to the Council on new long-term energy objectives for the Community[1]. The study has also been seen in Member States as a useful basis for comparative analysis, setting national projections in a wider context. Aside from its specific results, the work on **Energy 2000** has demonstrated the potential – both human and technical – developed within the Commission for the implementation of large-scale cross-country analytical exercises. The maintenance of such a capability, enabling the Commission to respond rapidly to a constantly changing external environment is of considerable importance.

The progress we have made in this field would not have been possible without close collaboration between experts in the development of models, energy economists and those responsible for advising on energy policy. All of them benefitted also from the help of Commission specialists in more general economic analysis.

In the preparation of the study the Commission staff were also able to develop very useful contacts with experts from national administrations and numerous energy companies. The informal opinions and advice given by these experts at various stages of the work have been if immense value.

We as Directors-General in the Commission are proud of this collaborative effort and we add our thanks to those of the authors themselves. Other products will no doubt follow from the team work that has been launched, casting new light in their turn on key energy questions facing the European Community.

Christopher AUDLAND   Paolo FASELLA
Director-General for Energy   Director-General for Science,
                                    Research, and Development

Commission of the European Communities

----

[1] COM(85)245 final of 28 May 1985.

ACKNOWLEDGEMENTS

Energy 2000 would not have been completed without extensive teamwork. A large number of economists and engineers played a part, within a broad framework of collaboration established between the Directorate-General for Science, Research and Development (DG XII) and the Directorate-General for Energy (DG XVII).

Particular thanks are due to Manuel Camos and Eric Donni, together with Georges Bingen, Bernard Funck and Jean-Marie Postiaux, for their analytical contributions to the study, and to Arnold De Backer and Claude Thonet for their data-processing support.

The authors would also wish to take this opportunity of thanking their colleagues from other parts of the Commission, notably the Directorate-General for Economic and Financial Affairs (DG II), the Forecasting and Assessment in Science and Technology (FAST) Group and the Statistical Office of the European Communities (SOEC) who gave valuable advice on specific aspects of the work. We also join our Directors-General in thanking the experts from national administrations and from the energy industries who helped to ensure that the study was not produced in an ivory tower.

Such a large undertaking would not have been successful without the commitment of senior management in the Commission, notably of David Hywel Davies, Deputy Director-General in DG XII, Directors Clive Jones and Albert Strub and Heads of Division Kevin Leydon and Ernst Roemberg. All of them gave their intellectual and moral support both in the preparation of the study itself and in ensuring a wide dissemination of its results throughout the Community.

The text of the report would, however, never have been completed without the patience, intelligence and dedication of two Commission secretaries, Sheila Reynolds and Anne-Marie Lits. The authors offer their particular thanks to them and to Jacques Blain for his work on the graphical presentation.

Energy 2000 has been produced independently by the Commission staff. They accept the full responsibility for both the assumptions used and the results. The advice given by outside experts - from national administrations and from the energy industries - was offered on an informal basis and the results do not of course commit their own organisations in any way whatever.

Jean-François GUILMOT   Pierre VALETTE
David McGLUE   Christian WAETERLOOS

S U M M A R Y

I. INTRODUCTION

1. During 1983 the Commission staff made an internal study of scenarios for the evolution of the Community energy markets to 2000[1] using the mathematical models developed under two Community research programmes on Energy Systems Analysis.

Three socio-economic scenarios were studied, representing three theoretical sets of developments for the Community economy up to the end of the century:

° a scenario called "Open Competition" which presupposed full-scale economic liberalism minimising intervention by public authorities;

° a scenario dubbed "Cooperation" based on a stimulus to world economic growth as a result of a successful North-South dialogue;

° a third scenario - "Europe" - which presupposed the primacy of European solutions to both economic and energy problems.

These rather exaggerated scenarios enabled the outer limits to the Community energy markets in 2000 to be explored and the problem areas and constraints to be identified. But in view of the rather theoretical nature of the exercise it was judged thereafter to be useful to define a fourth scenario of a more realistic nature as a basis for reflections within the Commission on Community energy policy in the medium and long-term, and specifically as a background to the definition of new Community energy policy objectives.

The study that follows is centred on that fourth scenario (termed henceforth the "reference projection"). It uses it as the basis on which to examine, through a series of complementary studies, the main uncertainties affecting the energy markets and the effects of different assumptions with regard to the development of energy prices, economic growth and sectoral energy policies, and so on.

2. Following a brief description of the methodology, this summary chapter describes the hypotheses adopted to construct the reference projection and the results obtained for energy supply and demand in the European Community (EUR-10) up to the year 2000[2].

----

[1] An internal staff paper outlining the main assumptions, methodology and results was given a limited distribution under the title "Scenarios to the year 2000: a prospective study of the energy market in the European Community".

[2] A preliminary version of this summary was published by the Commission

It outlines the main features of the reference projection for each of the 10 Member States and then sets the results in a world context. Finally it examines the effects of changing some of the key assumptions.

II. METHOD

3.   The exploratory study of the energy market up to the year 2000 was carried out by a joint working party from the European Commission's Directorate-General for Science, Research and Development and from the Directorate-General for Energy, with the support of experts from other services. It was carried out in six stages.

4.   The **first stage** focussed on the choice, as a general framework for the study, of a socio-economic scenario for the Community as a whole which reflected a broad consensus in the group on the probable development of the European economy and society in the latter part of the century[3]. The group drew on previous work under the FAST multiannual research programme for the purposes of defining this scenario and describing its main parameters in qualitative terms.

The **second stage** involved the specification of the relevant socio-economic parameters for each of the Member States and their translation into quantitative terms. In carrying out this phase of the study, the group made extensive use of the scenario study referred to earlier made with the aid of an informal multidisciplinary working party[4]. The results of these first two stages are contained in Part III below.

During the **third stage**, long-term projections of energy supply and demand in the Member States were made using mathematical models developed within the framework of the Community's programme of research on energy systems analysis[5]. Three sets of models were used.

The first, known as MEDEE-3, is techno-economic in nature, and is designed to simulate energy demand at the final consumption level: industry, transport, residential and tertiary sectors.

----

staff as a staff paper (SEC(85)324, February 1985). Some of the figures used here have been updated since that paper was published.

[3]"Forecasting and Assessment in the field of Science and Technology" (XII-A-1).

[4] The informal inter-departmental group included specialists in economics, industry, agriculture, transport, research, forecasting and energy. It held several meetings from May to October 1982.

[5] Energy Systems Analysis and Strategic Studies (XII-E-5).

The second, known as EFOM-12-C, uses linear programming to optimize the primary energy supply needed to meet a given level of final demand.

The third model, EURECA, established the necessary consistency between the economic variables, the price of imported oil and the other variables relating to the energy markets.

The **fourth stage** involved analysing the national results obtained and comparing them with the most recent national projections. The national administrations responsible for energy, together with several energy companies were consulted informally. The national results were then aggregated to arrive at a reference projection for the Community.

During the **fifth stage**, alternatives to the Community reference projection were studied. The aim of these variants was to assess the magnitude of the uncertainties affecting the energy market and to ascertain the sensitivity of the results to possible changes in the main assumptions.

The realism of the results obtained for the Community was analysed in the **sixth stage**, by setting these energy projections in the wider world context. To this end, energy supply and demand prospects in the various regions of the world up to the year 2000 were studied.

III. A REFERENCE PROJECTION FOR THE EUROPEAN COMMUNITY: MAIN ASSUMPTIONS

5. The reference scenario studied is based on the assumption of no major change to existing economic systems in the period under review. The economic framework assumes a gradual resumption of growth, but at a slower rate than in the sixties. It also assumes that the restructuring of industry will lead to a relative reduction in the share of basic industries.

Taking account of the analysis made by the Commission and the Council of the energy policies of each Member State,[6] the scenario assumes that as far as energy is concerned the Community's objectives set for 1990 will be achieved including, specifically, the completion of large programmes of construction of nuclear and solid-fuel fired electricity generating stations. It also assumes greater diversification of supply sources, both in terms of type of energy and origin.

----

[6] COM(84)87, COM(84)88 final, and COM(84)693 final.

(i)   General economic framework

6.   The annual growth rate of GDP in the Community dropped from 4.7% on average for the period 1960-1973 to 1.7% in the period 1974-1983.

The economic framework on which the reference projection is based assumes an average growth rate for the Community of 2.6% per annum for the period 1984-2000, resulting from a growth rate of 2.4% per annum between 1984 and 1990 and an increase to 2.8% per annum between 1991 and 2000.

This trend is assumed to be accompanied by a relatively stable share for private consumption and a reduction in public consumption in favour of investment. It is also supposed that there will be a gradual return to a positive trade balance for the Community, although the surplus will be very small. The ECU is expected to appreciate slightly against the US dollar from 1985.

7.   The hypotheses adopted concerning the structural development of GDP envisage an increase in the relative significance of services, of which the share is predicted to grow from 55% in 1980 to 59% in 2000. Consequently, there would be a decline in the shares of agriculture (-1%), the building trade (-1%) and industry (-2%).

It is assumed that the industrial restructuring which began at the end of the 1970s will lead in the year 2000 to a reduction in the share of the energy-intensive intermediate goods industries and primary processing industries, and a corresponding increase in capital goods industries. The share of the consumer goods industries would also fall.

Trend in the components and use of GDP

EUR-10	1970	1980	1983	1990	2000
Components of GDP (in %)					
Agriculture	4.8	4.3	4.3	4.1	3.4
- Energy-intensive[7] industries	10.2	9.2	9.0	8.6	7.7
- Other industries	26.0	25.3	25.3	25.2	24.9
- Building, public works	8.3	6.3	5.9	5.6	5.4
- Services	50.7	54.9	55.5	56.5	58.6
Use of GDP (in %)					
- Private consumption	60.0	63.2	64.0	63.8	63.8
- GFCF	23.7	20.8	19.5	20.8	21.3
- Public consumption	16.3	16.0	16.5	15.4	14.9

Capital Formation? (handwritten annotation next to "- GFCF")

----
[7] Metals and non-metallic minerals, pulp and paper, chemicals.

8. The population of the Community of Ten increased by 0.4% per annum on average over the last fifteen years. It now stands at 272 million inhabitants.

The divergent demographic trends between the countries of the north and the south of the Community should lead, in 2000, to a total population of around 280 million inhabitants.

(ii) Energy context

9. For the initial period, up to 1986-1987, it is assumed that the **average crude oil supply cost** to the Community, expressed in US dollars, will continue to fall in nominal terms. The gradual revival of world oil demand could then lead to a recovery in the international price of oil which would stabilize initially in real terms, and then increase by approximately 2.5% per annum from 1990 to reach $35 per barrel[8] in 2000. This assumption makes allowance both for trends in world oil demand, in particular on the part of the developing countries, and the revenue requirements of the oil exporting countries.

10. As − on this assumption − oil would continue to occupy a leading place in world energy supplies (approximately one-third), it would maintain its "guide price" function. The **prices of natural gas** would at least maintain their present advantage over oil products and the price differential might even increase under certain conditions. For the reference projection, however, it is assumed that the costs of production of imported coal will rise more rapidly than prices in general, which would lead to a slight reduction in the differentials between coal prices and those of oil products on the Community markets. **Electric power** would benefit from a somewhat more favourable trend in consumer prices than that for other fuels, owing to the advantages in production costs resulting from the penetration of nuclear power and increased recourse to solid fuels.

11. In general, the presumed trend in relative prices of the various energy sources should preserve or reinforce the existing advantages of competing energy sources vis-à-vis oil products. Under these conditions, the **replacement of oil** could be expected to continue in the long run at the same rate at least as that of equipment renewal.

12. The general climate should also remain favourable to **energy saving,** mainly because of the favourable effect of new investment and because of the long-term trend towards an increase in real prices, even if that increase would be much less pronounced than that experienced since the first oil crisis.

---
[8] Value in real 1983 terms; in money of the day, this price would correspond to the equivalent of $80/bbl in 2000.

The changing structure of industry at the expense of energy-intensive industries should reduce demand for energy per unit of value-added. An economic climate more favourable to investment than in previous years should accelerate the introduction of new industrial processes and the renewal of household equipment, resulting in a substantial improvement in efficiency. Further technical progress in energy equipment should also help to improve energy efficiency.

IV. A REFERENCE PROJECTION OF FINAL ENERGY DEMAND

(i) Background

13. Fast economic growth coupled with low energy prices caused final energy consumption to increase by an average of 5% per annum in the period **1960-1973.** This steady growth in consumption in all sectors (industry, transport, homes, non-energy uses) was chiefly sustained by oil products (+9% per annum), which covered the whole increase in the volume of demand.

Between **1973 and 1983,** on the other hand, energy consumption stabilized and then fell in all sectors of final demand except transport. Oil was replaced to a considerable extent by natural gas and electricity.

(ii) Overall projections to the year 2000

14. Given the assumptions made about the rate of economic growth for the period 1984-2000, the structural changes in the economy, energy prices and rational use of energy, **final energy demand** will increase slowly at an average around 0.9% per annum between now and the end of the century.

This rate of growth will be faster at the end of the 1980s (+1.3% per annum) than in the 1990s (+0.7% per annum). This is based on the assumption that the Community will emerge in the next few years from a period of economic recession and experience several years of falling prices for oil and energy in general. From 1990 onwards, however, the full effects of investment in rational use of energy and the impact of rising real energy prices should gradually make themselves felt.

This gradual slowing down in the growth of energy requirements should above all be apparent in the residential and tertiary sector, where energy requirements should gradually move towards saturation point. The energy requirements of the transport sector, on the other hand, are expected to increase steadily, although at a progressively slower rate than in the past. The same would apply to industrial demand.

The final energy intensity of the Community economy[9] should therefore

---

[9] The final energy intensity of the economy is calculated by dividing

improve by one quarter over the projection period, falling from 0.49 toe/10^3 ECU in 1983 to 0.46 in 1990 and finally to 0.38 in 2000.

Trend in final energy consumption, by sector
(in million toe)

EUR-10	1973	(%)	1980	(%)	1983	(%)	1990	(%)	2000	(%)
Industry	247.9	(35)	227.0	(32)	186.4	(29)	216	(30)	234	(31)
Transport	128.2	(18)	153.6	(22)	155.6	(24)	169	(24)	182	(24)
Residential and tertiary	264.9	(37)	265.8	(38)	252.6	(38)	267	(37)	274	(36)
Non-energy use	0.3	(10)	60.0	(8)	57.4	(9)	63	(9)	72	(9)
Total	711.3	(100)	706.4	(100)	652.0	(100)	714	(100)	762	(100)

15. Following the trend of the last few years, natural gas and electricity should meet most of the increase in demand, with their share in final energy consumption increasing accordingly.

District heating and renewable sources of energy are also likely to develop significantly **in relative terms.**

Solid fuels should maintain their current share in final consumption, whereas that of oil products should decrease substantially, with consumption in volume flattening out around present levels, i.e. 35 mtoe less than in 1980.

Trend in final energy consumption, by sources
(in million toe)

EUR-10	1973	(%)	1980	(%)	1983	(%)	1990	(%)	2000	(%)
Solid fuels	85.4	(12)	63.1	(9)	55.3	(8)	58	(8)	61	(8)
Oil products	439.3	(62)	395.1	(56)	351.2	(54)	363	(51)	359	(47)
Gas	107.4	(15)	151.5	(21)	147.2	(22)	169	(23)	186	(24)
Electricity	76.0	(10.5)	92.8	(13)	94.6	(15)	113	(16)	138	(18)
Heat and others[10]	3.2	(0.5)	3.9	(1)	3.7	(1)	11	(2)	18	(3)
Total	711.3	(100)	706.4	(100)	652.0	(100)	714	(100)	762	(100)

(iii)   Sectoral projections

16. Three principal factors should influence energy consumption in industry in the Community between now and the year 2000:
----

final energy consumption including non-energy uses, expressed in toe, by GDP, expressed in ECU at 1975 prices and exchange rates.

[10] Including 2.5 mtoe of heat in 1983, 6 mtoe in 1990 and 10 mtoe in 2000.

- changes in **industrial structure,** resulting in an increase in the share of light industries and capital goods industries at the expense of the energy-intensive industries, and thus in a reorientation of production towards products with a higher added value;

- the shift towards **new processes** which make more efficient use of energy;

- the **modernization** of existing production plant, which should improve its energy performance and, in many cases lead to substitution between types of energy.

These three complementary factors would have the effect of gradually reducing specific energy consumption in industry from 0.44 toe/10^3 ECU of value added in industry (in real terms) in 1983 to 0.41 in 1990 and 0.36 in the year 2000.

Although industrial activity is expected to grow by an average of 2.3% per annum, energy consumption in this sector will increase annually by only 1.3% until the end of the century. Oil consumption will rise in volume terms until 1990 and then return to about its 1983 level in the year 2000, whereas consumption of steam coal, natural gas and, above all, electricity will increase slightly throughout the whole period.

Trend in energy consumption in industry
(in million toe)

EUR-10	1973	1980	1983	1990	2000
Solid fuels	45.6	38.4	35.7	40	46
Oil products	104.7	77.6	49.3	57	52
Gas	57.9	66.8	59.6	66	73
Electricity	38.6	42.8	40.6	49	58
Heat and others	1.1	1.4	1.2	4	5
Total	247.9	227.0	186.4	216	234

17. Activity in the **transport** sector is traditionally linked, as far as the carriage of goods is concerned, to the general pattern of economic activity. However, the link between economic growth and energy demand for this mode of transport is likely to weaken progressively as a result of the structural change in GDP in favour of services and of technological progress.

Passenger transport is linked to the specific features of social organization (housing characteristics, distance from workplace, leisure, household income patterns). With regard to its modal structure, passenger transport could undergo two important developments: a slight increase in the share of public transport and, parallel to this, an increase in the number of private vehicles - but with a lower average annual distance travelled per vehicle.

Technological advances in vehicle design and technical improvements to engines should, moreover, lead to increased efficiency of the transport sector as a whole, leading to a decrease in energy consumption per monetary unit of GDP from 0.12 toe/10^3 ECU (real terms) in 1983 to 0.11 in 1990 and 0.09 in 2000.

On the basis of the assumed rate of economic growth, the combination of the various factors listed above should result in a moderate growth in energy consumption in the transport sector of 0.9% per annum between 1983 and 2000. This would involve almost exclusively increased consumption of motor fuels where there will be a substantial increase in the relative share of diesel.

Trend in energy consumption in the transport sector
(in million toe)

EUR-10	1973	1980	1983	1990	2000
Oil products	125.1	150.7	152.8	166	178
Electricity	2.0	2.4	2.4	3	4
Others[11]	1.1	0.5	0.4	0	0
Total	128.2	153.6	155.6	169	182

18. 70% of energy consumption in the residential and tertiary sectors is accounted for by space heating requirements.

Given the rate of renewal or renovation of buildings assumed in the reference projection, together with improvements in energy efficiency and technical characteristics of heating equipment, there could be a virtual stagnation of the total energy required for space heating until 1990, followed by a reduction in the subsequent ten years.

The other uses of energy in households, craft industry, trade and services include the production of domestic hot water, cooking and specific uses of electricity for motive power, household appliances and lighting.

Energy consumption for each of these uses is likely to continue to increase until the end of the century. The increase in the case of specific uses of electricity is likely to be relatively large at around 3% per annum between 1983 and 2000, even when allowance is made for expected technological advances.

On the whole, energy consumption in the residential and tertiary sector should continue to grow by an average of 0.6% per annum until the year 2000.

Oil products and solid fuels will continue to be replaced by natural gas, electricity and district heating from heating stations or mixed power stations (coal- or gas-fired) or from recovered steam.

---

[11] Solid fuels and natural gas.

Trend in energy consumption in the residential and tertiary sector
(in million toe)

EUR-10	1973	1980	1983	1990	2000
Solid fuels	35.4	21.8	18.2	16	13
Oil products	148.8	117.4	101.3	90	71
Gas	43.2	76.5	79.0	92	100
Electricity	35.4	47.6	51.6	61	77
Heat and others	2.1	2.5	2.5	8	13
Total	264.9	265.8	252.6	267	274

19.   Products for non-energy uses can be classed in two groups: oils,
lubricants, bitumens and waxes on the one hand, and raw materials for
the petrochemicals industry on the other. There will be a slight
increase in demand for the two groups of product until the end of the
century, with no change in the overriding importance of oil products.

Overall, consumption of all these products should grow by 1.3% per
annum until 2000.

Trend in consumption of products for non-energy uses
(in million toe)

EUR-10	1973	1980	1983	1990	2000
Solid fuels	3.4	2.7	1.3	2	2
Oil products	60.7	49.4	47.8	51	58
Gas	6.2	7.9	8.3	10	12
Total	70.3	60.0	57.4	63	72

V.   GROSS ENERGY CONSUMPTION TO THE YEAR 2000

20.   Gross energy consumption in the Community increased by an average of
4.7% per annum between 1963 and 1973. It then followed an erratic
development before reaching a level in 1980 close to that in 1973.

Gross energy consumption[12] as a whole in the Community should
increase on average by only 1.7% per annum until 1990, and then by 1%
per annum during the following decade, which is equivalent to an
average increase of 1.3% per annum between now and the year 2000.

The moderate long-term increase in energy demand in the Community not
only reflects the moderate economic growth rate assumed, and changes
in economic structure, but above all the results of sustained and
effective policies for rational use of energy.

----

[12] Including bunkers.

21.   Between 1973 and 1983, the intensity of primary energy use in the Community improved by about 20%. A further improvement of the same magnitude should be achieved between 1983 and 2000.

Thus, on the basis[13] of the assumptions made, the energy ratio of the Community economy[13] should decline slightly, from 0.67 toe/10^3 ECU in 1983 to 0.64 in 1990, followed by a further sharp drop to 0.54 in 2000.

This trend would be the consequence of structural changes, future investment in new, less energy-intensive processes and modernization of equipment leading to improved energy performance.

22.   Taking into account the supply and demand conditions of the electricity sector (see paragraphs 28-30 below), the trend in the relative shares of the various sources of energy in gross consumption in the Community would be as follows:

- a stable share for solid fuels;

- substantial reduction in the share of oil to less than 40% in 2000;

- considerable increase in the role of nuclear power which should meet 19% of energy requirements at the end of the century;

- slight increase until 1990, followed by a reduction in the relative share of natural gas. (This is in contrast to the evolution in gas' share of final energy consumption which is expected to grow steadily. The difference results from the expected fall in gas consumption in power stations).

Trend in gross energy consumption in the Community (including bunkers) - (in million toe)

EUR-10	1973	(%)	1980	(%)	1983	(%)	1990	(%)	2000	(%)
Solid fuels	222.0	(23)	222.7	(23)	212.2	(23)	242	(23)	264	(23)
Oil	601.3	(62)	520.0	(54)	438.3	(48)	441	(43)	439	(39)
Natural gas	115.8	(12)	169.3	(17)	165.2	(18)	190	(18)	196	(17)
Nuclear energy	17.7	(2)	42.7	(4)	76.1	(9)	145	(14)	215	(19)
Hydro, geothermal, other	11.0	(1)	15.4	(2)	15.5	(2)	12	(2)	21	(2)
Total	967.8	(100)	970.1	(100)	907.3	(100)	1034	(100)	1136	(100)

----

[13] The energy ratio of the economy is calculated by dividing internal primary energy consumption (without bunkers) by GDP in real terms (1975 prices and exchange rates).

VI. AN ENERGY SUPPLY SCENARIO CORRESPONDING TO THE REFERENCE PROJECTIONS

(a) Historical background

23. Between **1963 and 1973,** internal energy production in the Community
stabilized at approximately 350 million toe, with the increase in the
production of natural gas balancing the reduction in coal production.
The increase in energy demand was thus covered entirely by higher
imports, which rose from 260 million toe in 1963 to 620 million toe
in 1973. Oil accounted for 95% of these imports.

These trends were completely reversed **after 1973,** with the result
that internal production in 1983 stood at 516 million toe and net
imports at 378 million toe, of which oil then only constituted
three-quarters.

Community dependence on external sources of supply thus rose from 43%
to 64% during the period 1963-1973 and then dropped to 42% ten years
later.

(b) Energy production in the Community

24. Nuclear energy is likely to be the only energy source whose
production level will increase significantly in the Community by
2000. According to the reference projection, production of solid
fuels should remain static at around the 1983 level. Production of
oil and natural gas should decline gradually on the basis of fields
being exploited and those likely to come into production.

Internal production of energy in the Community
(in million toe)

EUR-10	1973	(%)	1980	(%)	1983	(%)	1990	(%)	2000	(%)
Coal	171.2	(48)	153.3	(33)	143.0	(28)	139.0	(25)	137	(22)
Lignite and peat	26.5	(8)	31.8	(7)	31.0	(6)	36.0	(6)	35	(6)
Oil	13.1	(4)	91.1	(20)	132.5	(26)	111.0	(20)	108	(17)
Natural gas	112.2	(32)	129.2	(28)	119.8	(23)	115.0	(21)	108	(17)
Nuclear energy	17.7	(5)	42.7	(9)	76.1	(15)	145.0	(26)	215	(35)
Hydroelectric and geothermal energy	9.1	(3)	12.3	(3)	12.0	(2)	13.0	(2)	14	(2)
Others & renewable energy sources	1.2	(0)	1.7	(0)	1.7	(0)	3.0	(0)	7	(1)
Total	353.2	(100)	462.1	(100)	516.1	(100)	563.0	(100)	625	(100)

The above figures for energy production in the Community in 1990 are generally very close to the projections communicated by the Member States[14] during the last "Review of national energy programmes".

The "reference projection" figures **for 2000** were obtained by the definition of a set of plausible assumptions about supply availabilities. Nevertheless these assumptions contain three major uncertainties:

- a question mark about the addition of new, competitive coal production capacities in the Community, without which production levels would decline considerably in the 90s;

- the level of new investments in the nuclear sector, without which the expected increase in nuclear's contribution 1990-2000 would not take place;

- the possibility of discovering one or more major oil fields or natural gas fields in the Community.

(c) Energy imports from third countries

25. Energy imports into the Community should return to 1979 levels by 2000, i.e. approximately 530 million toe. After having reached their lowest level (378 million toe) in 1983, **imports of coal, oil and gas would thus increase steadily until 2000.**

Net energy imports into the Community
(in million toe)

EUR-10	1973 (%)	1980 (%)	1983 (%)	1990 (%)	2000 (%)
Solid fuels	19.0 (3)	47.3 (9)	39.1 (11)	67 (14)	92 (18)
Oil	596.2 (96)	437.9 (83)	288.8 (76)	330 (70)	331 (65)
Natural gas	4.0 (1)	40.6 (8)	48.2 (13)	74 (16)	88 (17)
Electricity	0.7 (0)	1.4 (0)	1.8 (0)	– –	
Total	619.9(100)	527.2 (100)	377.9 (100)	471 (100)	511 (100)

Attempts will probably be made in two respects to diversify imports until the end of the century. On the one hand, there will be a shift in the respective shares of oil and the other imported fuels in total imports which would develop from a ratio of 83%:17% in 1980 to 65%:35% in 2000. On the other, there will be greater diversification in the geographical source of these imported supplies.

---

[14] See COM(84)88 of 29 February 1984.

26.   The Community's dependence on imported oil for its energy supply
      should lessen considerably during the next 15 years. It was
      successfully reduced from 62% in 1973 to 32% in 1983. This figure
      should be reduced to less than 30% in 2000, if the basic assumptions
      made for the reference projection are borne out in reality.

      On the other hand, dependence on imported gas should continue to rise
      until 2000. From zero in 1973, it reached 5% in 1983 and is likely to
      grow to 8% in 2000 because of the rise in consumption combined with a
      drop in internal production.

      For the same reasons, dependence on imported coal should also
      increase, rising from 2% in 1973 to 4% in 1983 and 8% in the year
      2000.

      Consequently total dependence on imported energy should increase from
      less than 42% in 1983 to around 45% in 2000.

      (d)   Electricity production

27.   The reference projection for final energy consumption puts final
      energy demand in 2000 at 763 million toe, 138 million toe of which
      would be accounted for by electricity. Final electricity consumption
      would thus increase by 2.2% per annum between 1983 and 2000.
      Electricity would therefore cover about one third of the increase in
      final energy demand. Its share would grow from 15% in 1983 to 18% in
      2000, whereas in 1973 this share was barely more than 10%.

      The increased penetration of electricity reflects an expected
      widening of markets for the specific uses of electricity, and in
      certain countries also the increased competitiveness of electricity
      for heating purposes. In the reference case, however, this
      penetration would be slowed down to a certain extent by a general
      improvement in the efficiency of electrical equipment.

      It is nevertheless necessary to emphasize the major uncertainties
      involved in the development of electricity demand. The growth rate
      derived from the country-by-country analysis would imply a reversal
      of the past trends in the ratio between economic growth and
      electricity demand. This has always tended to be equal to or greater
      than one, whereas under the reference scenario it will not exceed
      0.85 between now and the end of the century.

28.   The generating plant needed to satisfy projected electricity demand
      will probably change considerably by the end of the century, with
      nuclear power stations producing about 43% of the electricity and 39%
      being generated by power stations fired with solid fuels. An
      investment programme providing for 100 Gw from nuclear power and
      40 Gw from power stations burning solid fuels between 1983-2000 would
      be required to correspond to this production profile.

Trend in total net electricity production, by energy sources
(in TWh)

EUR-10	1973	(%)	1980	(%)	1983	(%)	1990	(%)	2000	(%)
Coal	301.1	(30)	412.8	(34)	423.3	(34)	513	(34)	599	(32)
Lignite and peat	78.1	(8)	98.1	(8)	106.3	(9)	118	(8)	134	(7)
Oil products	312.2	(32)	264.8	(22)	158.5	(13)	83	(5)	71	(4)
Natural gas	100.5	(10)	107.6	(9)	95.7	(8)	101	(7)	56	(3)
Derived gas	23.8	(2)	21.1	(2)	16.1	(1)	18	(1)	16	(1)
Nuclear energy	53.5	(5)	149.4	(12)	275.0	(22)	534	(35)	792	(43)
Hydroelectric and geothermal energy	112.9	(12)	148.8	(12)	147.4	(12)	150	(10)	165	(9)
Others	4.6	(1)	6.4	(1)	7.0	(1)	8	(1)	14	(1)
Total	986.7	(100)	1209.0	(100)	1229.3	(100)	1523	(100)	1847	(100)

29. If optimum use is to be made of electricity generating capacity in
     the Community, nuclear and coal-fired power stations should, from
     1990, normally cover the base and middle load, with oil- and
     gas-fired plant being used to cover the remainder of the load and
     peak demand.

     The following table shows the corresponding trend in energy
     consumption in the electricity generating stations in the Community.

Trend in energy consumption for electricity production
(in million toe)

EUR-10	1973	1980	1983	1990	2000
Solid fuels	101.3	130.1	133.9	160	178
Oil products	75.0	60.9	36.9	20	18
Gas (natural and derived)	30.6	31.3	26.2	27	15
Nuclear energy [15]	17.7	42.7	76.1	145	215
Hydroelectric and geo-thermal energy [16]	9.1	12.3	12.0	13	14
Other products	1.2	1.7	1.7	2	4
Total	234.9	279.0	285.8	367	444

---

[15] Fission heat is considered to be of primary origin using the SOEC
method.

[16] Electricity derived from hydroelectric or geothermal energy is counted
as primary energy using the SOEC method.

VII. MAIN ASSUMPTIONS MADE IN THE REFERENCE PROJECTION, COUNTRY BY COUNTRY

30. The reference projection outlined above refers to the Community as a whole, but was calculated by adding together individual projections made separately for each Community country.

   These national projections were all based on a comparable set of hypotheses derived from the general social and economic framework chosen for the Community as a whole. Consequently, certain aspects of these hypotheses may diverge from those adopted in, for instance, the latest energy projections used by individual national administrations.

   This section briefly sums up the results obtained for the reference projection, country by country.

(a) Belgium

31. Assuming that the Belgian economy grows by an average of 2% a year between 1983 and 2000, total energy demand in Belgium can be expected to increase by 1.3% a year, from 42.7 million toe in 1983 to 53 million toe in 2000.

   The primary energy ratio of the Belgian economy has already fallen by roughly 30% since 1973, largely as a result of the restructuring of Belgian industry, and can be expected to fall by a further 12% or so by the end of the century, from 16.5 kgoe/10^3 Bfr[17] in 1983 to 14.5 Bfr in 2000.

   Final electricity consumption is expected to rise by 2.1% a year between 1983 and 2000, by which time electricity should cover 15% of final energy demand, compared with 13% in 1983.

Gross primary energy consumption in Belgium (including bunkers)
(in million toe)

	1973	1980	1983	1990	2000
Solid fuels	11.3	11.1	9.3	9.4	10.1
Oil	30.3	25.2	20.1	20.9	21.5
Natural gas	7.2	9.0	7.1	8.4	8.5
Nuclear energy	0	3.1	6.1	9.7	12.8
Hydroelectric power and others	-	-0.2	0.1	0.1	0.1
Total	48.8	48.2	42.7	48.5	53.1

---

[17] kgoe/10^3 Bfr = kilogram of oil equivalent per thousand Belgian francs at constant 1975 prices.

32. No increase is expected in the consumption of solid fuels before 1990.

    This is largely because of the decline in consumption in the steel industry (as a result of more efficient processes) and in power stations (as a result of increasing recourse to nuclear energy). On the other hand, the gradual conversion to coal by a number of large industrial energy consumers should offset this loss of market share.

    After 1990 a substantial increase in sales to the electricity industry coupled with the forecast expansion of combined heat and power production should ensure that the market share of solid fuels stabilizes, without, however, ever returning to 1980 levels.

    Natural gas consumption is also expected to grow steadily again to return to close to 1980 levels by 2000. Sales to the domestic and other sectors should rise as orders from power stations fall.

    Led by the expected economic recovery, oil consumption should increase slightly by 2000, though it will not approach 1980 or 1973 levels and indeed its share of gross domestic energy consumption should even decline from 47.1% in 1983 to 40.4% in 2000.

    Together, the increase in motor fuel sales and the levelling-off of sales to the domestic and other sectors will increase the share of light and medium distillates and reduce that of the heavier products.

33. Nuclear energy is expected to double its share by 2000, reducing Belgium's dependence on imported energy supplies from 85% in 1980 to 67% by 2000.

Primary energy supplies in Belgium
(in million toe)

	1973	1980	1983	1990	2000
Domestic production	5.9	7.8	10.8	14.0	16.9
of which: nuclear energy	0	3.1	6.1	9.7	12.8
Net imports	42.9	41.4	31.5	34.5	36.2
of which: oil	30.5	25.6	20.0	20.9	21.5
natural gas	7.1	9.0	7.2	8.4	8.5
Stock change	-	-1;0	+0.4	..	..
Total	48.8	48.2	42;7	48.5	53.1

Belgium's coal output is assumed to hold steady at around 4 million toe a year up to the end of the century.

Oil imports are expected to rise by roughly 8% and natural gas imports by 18% between 1983 and 2000.

These extra imports should cover roughly 35% of the projected extra demand for energy, with nuclear energy coping with the remainder.

(b) **Denmark**

34. Between 1983 and 2000 gross energy consumption in Denmark is expected to grow by an average of 1.7% a year and GDP by 2.4% a year.

    This would bring the primary energy ratio of the Danish economy down from 71 kgoe/10^3 Dkr[18] in 1983 to 64 kgoe/10^3 Dkr in 2000. After the 25% reduction between 1973 and 1982, Denmark's primary energy intensity is unlikely to fall by much more than 10% between now and the end of the century. At first sight this may appear modest, but it reflects the fact that over the next few years Denmark will probably put the accent on flexibility and diversification, favouring district heating, and, to a lesser extent, electricity.

    As a result, heat consumption in Denmark can be expected to rise by 3.0% a year by the end of the century, compared with 2.0% a year for final energy consumption. This would bring electricity's share of final energy demand up from 15.5% in 1983 to 17.7% in 2000.

Gross primary energy consumption in Denmark (including bunkers)
(in million toe)

	1973	1980	1983	1990	2000
Solid fuels	2.3	5.9	5.4	8.4	10.2
Oil	17.9	13.7	10.8	10.4	9.8
Natural gas	-	-	-	1.7	1.9
Nuclear energy	-	-	-	-	-
Hydroelectric power and others	-0.0	0.1	0.4	0.2	0.4
Total	20.2	19.7	16.6	20.7	22.3

35. In the medium term, total consumption of oil products is likely to fall as the moves to replace oil products continue in all but premium applications such as transport. In this case the pattern of consumption will shift towards lighter products at the expense of medium fractions, and in particular of heating oil.

    Heating oil is already facing competition from natural gas which made its first contribution to Denmark's energy supply in 1984. Natural gas can be expected to increase its share as output from the Tyra field in the Danish section of the North Sea is stepped up.

    In addition, the gradual spread of district heating networks, which are normally coal-fired, should also speed up the moves to replace gas oil in the domestic and equivalent sectors.

    The anticipated increase in demand for electricity, most of it generated from solid fuels, should considerably increase coal's share in Denmark's energy supply. Almost 85% of the extra demand for energy between 1983 and 2000 is likely to be covered by solid fuels, with natural gas accounting for the rest.

----

[18] kgoe/10^3 Dkr = kilogram of oil equivalent per thousand Danish kroner at constant 1975 prices.

36.   The exploitation of Denmark's indigenous oil and gas resources over
the next few years should reduce Denmark's dependence on imported
energy supplies significantly (from 87% in 1983 to 69% in 2000).
Output should rise to 6.5 million toe by 1990 and hold steady at that
level thereafter.   Plans are being made to conclude medium-term
contracts to export roughly 0.7 million toe of natural gas a year to
Germany and Sweden.

Primary energy supplies in Denmark
(in million toe)

	1973	1980	1983	1990	2000
Domestic production	0.1	0.3	2.2	6.5	6.9
of which: oil	0.1	0.3	2.2	4.0	4.0
natural gas	-	-	-	2.4	2.6
Net imports	20.3	19.4	14.2	14.2	15.4
of which: coal	2.0	6.1	5.4	8.4	10.2
oil	18.3	13.2	8.5	6.4	5.8
Stock change	-0.2	0.0	+0.2	..	..
Total	20.2	19.7	16.6	20.7	22.3

(c) **Germany**

37.   Although the economy is expected to grow by an average of 2.7% a year
between 1983 and 2000, total energy demand is unlikely to increase by
more than 0.8% a year, on average, over the same period.

This would give a 27% reduction in the primary energy ratio of the
German economy, down from 200 kgoe/10^3 DM[19] in 1983 to 146 in 2000,
after a reduction of almost 20% between 1973 and 1982. This is due to
a far-reaching restructuring of the economy on the one hand and
substantial energy savings on the other.

Final electricity consumption is expected to rise by an average of
1.8% a year between 1983 and 2000, thus raising electricity's share
of final energy consumption from 17% in 1983 to 19.5% in 2000.

Gross primary energy consumption in Germany (including bunkers)
(in million toe)

	1973	1980	1983	1990	2000
Solid fuels	83.2	82.7	81.5	83.8	88.4
Oil	149.8	131.7	110.5	111.9	104.3
Natural gas	27.0	44.6	39.6	43.3	42.1
Nuclear energy	3.0	11.1	16.5	34.1	50.3
Hydroelectric power and others	2.8	2.9	3.3	3.5	4.5
Total	265.8	273.0	251.4	276.6	289.6

----
[19] kgoe/10^3 DM = kilogram of oil equivalent per thousand German marks at
1975 prices.

38. Consumption of oil products is expected to remain relatively stable throughout the period, with a slight increase in sales of light products, and in particular of petrochemical cuts and kerosene, and a downturn in sales of medium products. Here the increase in diesel fuel deliveries will not be enough to offset the big fall in heating oil sales which should result from a more efficient use of heating oil and a switch to natural gas, district heating and electricity in the domestic and tertiary sectors.

    Consequently, demand for natural gas can be expected to return to close to the 1980 level and to remain there between 1990 and 2000. This presupposes increased sales to the domestic and tertiary sectors, consolidation of its share of the industrial market and a slight downturn in natural gas sales to power stations in the long run.

    Since the German government has long been encouraging moves to burn coal in power stations or for combined heat and power production, whether in industry or for district heating schemes, consumption of solid fuels seems likely to increase even more by 2000.

    The reference projection also assumes a substantial expansion of nuclear capacity, enabling nuclear energy to make an even greater contribution to Germany's overall energy balance and allowing electricity to capture new markets, particularly for heat end-uses.

39. This expansion of the nuclear programme combined with exploitation of Germany's indigenous solid fuel, oil and gas resources should make Germany slightly less dependent on imported supplies of energy, with the rate falling from 52% in 1983 to 48.4% in 2000.

Primary energy supplies in Germany
(in million toe)

	1973	1980	1983	1990	2000
Domestic production	119.2	121.4	120.6	137.8	149.4
of which: hard coal	69.1	62.2	58.4	56.3	52.7
brown coal	22.9	26.5	25.3	26.0	25.0
natural gas	15.0	14.3	13.6	13.8	13.8
nuclear energy	3.0	11.1	16.5	34.1	50.3
Net imports	147.4	157.0	128.4	138.8	140.2
of which: oil	144.6	131.2	102.7	106.9	100.3
natural gas	12.0	30.6	26.3	29.5	28.3
hard coal	-4.4	-2.2	-0.5	1.5	10.4
Stock change	- 0.8	- 5.4	2.4	..	..
Total	265.8	273.0	251.4	276.6	289.6

Imports of natural gas and oil products should remain relatively close to 1983 levels over the whole period, though imports of hard coal should increase considerably.

The expected increase in nuclear output should account for a very large share of the incremental demand for energy anticipated in Germany between 1983 and 2000.

(d) Greece

40. An average annual growth rate of 2.6% was assumed for the Greek economy throughout the period from 1983 to 2000, with gross energy consumption growing by 2.9%. The industrial development planned over the next fifteen years is the main reason why demand for energy is expected to grow faster than the economy as a whole. Together with Ireland, Greece is one of the two Community countries where manufacturing industry is expected to **increase** its share of GDP.

In the process the primary energy ratio of the Greek economy is likely to increase from 25 kgoe/10^3 Dr[20] in 1983 to 26.5 in 2000. This coefficient had already risen by roughly 5% between 1973 and 1983.

Final electricity consumption should rise by 3.6% a year up to the end of the century, from 16% of final energy consumption in 1983 to 18% in 2000.

Gross primary energy consumption in Greece (including bunkers)
(in million toe)

	1973	1980	1983	1990	2000
Solid fuels	2.2	3.4	4.8	8.0	10.0
of which: lignite	1.8	3.1	3.8	7.3	9.0
Oil	10.1	2.4	11.5	13.0	15.3
Natural gas	–	–	0.1	0.1	0.6
Hydroelectric power and other	0.2	0.3	0.3	0.5	0.9
Total	12.5	16.1	16.7	21.6	26.8

41. The most striking change in Greece's energy balance in the medium term is likely to be a major increase in consumption of solid fuels. This will be due partly to the large programme to exploit Greece's indigenous lignite resources and to convert Greece's power stations to lignite, and partly to the inroads that could be made by solid fuels in the industrial sector.

In the long term natural gas is expected to meet a very small proportion of energy requirements in the industrial and residential sectors.

----

[20] kgoe/10^3 Dr = kilogram of oil equivalent per thousand Greek drachmas at 1975 prices.

Output from hydroelectric and geothermal power stations is also expected to rise sharply by the end of the century without, however, becoming one of the major sources of supply for the Greek energy market.

One difference between Greece and the other Community countries is that a substantial increase in consumption of oil products will be needed to meet her growing energy requirements.

42. As Greece progressively turns to its indigenous energy resources (lignite, hydroelectricity and new and renewables) to generate its electricity it should be able to keep its energy dependence rate down to around 65% between 1983 and 2000 despite the sharp increase in its overall energy requirements.

Primary energy supplies in Greece
(in million toe)

	1973	1980	1983	1990	2000
Domestic production	2.0	3.4	5.4	8.2	9.9
of which: lignite	1.8	3.1	3.8	7.3	9.0
oil	-	-	1.2	0.3	-
Net imports	11.6	13.4	10.9	13.4	16.9
of which: oil	11.1	13.0	9.9	12.7	15.3
Stock exchange	-1.1	-0.7	0.4	..	..
Total	12.5	16.1	16.7	21.6	26.8

Greece will probably at least double its lignite output by the end of the century. At the same time its steam coal imports should hold steady at around 1 million toe.

Greece will also probably be producing small amounts of natural gas by the 1990s. Towards the end of the century, it will probably import natural gas to top up its own output - presumably from Algeria, the Soviet Union or even Italy.

Finally, for the next fifteen years imported oil should remain the largest single source of supply for the Greek energy market even though oil products' share of gross domestic energy consumption is expected to continue to decline, from 69% in 1983 to slightly over 57% in 2000.

(e) **France**

43. For the period 1983 to 2000, it has been assumed that total demand for energy in France will grow by an average of 1.6% a year and GDP by 2.6% a year.

This should make the French economy roughly 20% less energy-intensive by the end of the century, with its primary energy ratio dropping from 120 kgoe/10^3 FF[21] in 1983 to 100 in 2000, following the earlier 20% reduction between 1973 and 1982.

Final electricity demand is expected to increase by 3.2% a year up to 2000, lifting electricity's share of final energy consumption from 15% in 1983 to 23% in 2000.

Gross primary energy consumption in France (including bunkers)
(in million toe)

	1973	1980	1983	1990	2000
Solid fuels	28.7	31.1	25.2	23.6	27.9
Oil	129.2	113.0	89.6	84.2	80.1
Natural gas	13.6	21.6	22.4	26.5	26.9
Nuclear energy	4.5	16.3	37.4	74.3	93.4
Hydroelectric power and other	4.0	6.4	5.1	4.0	6.9
Total	180.0	188.4	179.7	212.6	235.2

44. The major feature of the French energy outlook over the next few years will be the growing share taken by nuclear energy at the expense of oil products in particular, with France's nuclear power capacity escalating from 28 MWe in 1983 to roughly 50 MWe in 1990 and to close to 70 MWe in 2000.

This rapid expansion of nuclear capacity should push down coal consumption by power stations until 1995 or so. Nonetheless, the spread of district heating and the gradual conversion of various industries to coal should help to offset in the longer-term this slackening of the total demand for coal.

The market for natural gas is likely to grow slightly throughout the 1980s, mainly in the residential and tertiary sector but also, to a lesser extent, in industry, so as to absorb the supplies available under import contracts already signed. After 1990 the natural gas market should stabilize.

These prospects for solid and gaseous fuels should progressively reduce consumption of oil products. These will be focussed increasingly on premium markets such as transport and non-energy applications.

This shift would give light products a larger share of total oil consumption.

---

[21] kgoe/10^3 = kilogram of oil equivalent per thousand French francs at 1975 prices.

45.   Even without any increase in domestic fossil fuel output, France's
      nuclear power programme alone should be enough to reduce France's
      dependence on imported energy supplies from 65% in 1983 to 50% in
      2000.

Primary energy supplies in France
(in million toe)

	1973	1980	1983	1990	2000
Domestic production	34.3	43.	63.4	97.7	118.2
of which: coal	16.4	11.7	10.8	8.2	8.0
nuclear	4.5	16.3	37.4	74.3	93.4
Net imports	145.9	149.1	110.5	114.9	117.0
of which· coal	9.9	20.0	11.8	14.6	19.1
oil	128.7	112.6	81.0	79.7	75.6
natural gas[22]	7.6	16.2	18.	23.6	23.9
electricity	-0.3	0.3	-12	-3.0	-1.6
Stock exchange	-0.2	-4.6	5.8	..	..
Total	180.0	188.4	179.7	212.6	235.2

The extra nuclear output should be enough to cover the entire
increase in energy requirements between 1983 and 2000.

Coal production in France is expected to fall as a result of the
rationalization of French mines, and a substantial increase in net
steam coal imports is expected.

Natural gas imports should rise by about a quarter by 1990 under the
contracts already concluded, but should hold steady thereafter at
around 24 million toe. Oil imports should remain unchanged until
1990. Thereafter another small reduction is possible.

France started to export electricity generated at its nuclear plants
in the early 1980s and can be expected to continue to do so until the
end of the century, with exports peaking in around 1990.

(f)   **Ireland**

46.   It was assumed for the study that the Irish economy would grow by an
      average of 3.5% a year between 1983 and 2000, with total energy
      demand growing by 2.5% a year over the same period. This would make
      the Irish economy 15% less energy-intensive by the end of the[23]
      century, with the primary energy ratio falling from 2 kgoe/£Irl in
      1983 to 1.7 in 2000, in addition to the 12.5% reduction between 1973
      and 1982.

      Final electricity consumption is expected to rise by 2.4% a year up
      until 2000. This would make Ireland the only country in the Community
      with electricity consumption growing slightly slower than energy

---

[22] A minus sign (-) indicates net exports.

[23] kgoe/£Irl = kilogram of oil equivalent per Irish pound at 1975 prices.

consumption and holding its share steady (at more or less 12%). This is primarily because natural gas is expected to make deep inroads into the industrial, residential and tertiary markets.

Gross primary energy consumption in Ireland (including bunkers)
(in million toe)

	1973	1980	1983	1990	2000
Solid fuels	1.4	2.0	2.0	3.3	4.3
of which: brown coal	0.7	1.1	1.0	1.1	1.3
oil	5.5	5.7	4.2	5.3	5.9
Natural gas	-	0.7	1.8	1.4	1.7
Hydroelectric power and others	0.1	0.1	0.1	0.2	0.3
Total	7.0	8.5	8.1	10.2	12.2

47.   This increase in energy consumption in Ireland should be accompanied by a greater diversification of supply, reducing the share of oil products in gross energy consumption to 48% in 2000, as against 52% in 1983.

The absolute level of consumption of oil products is, however, expected to increase, led by the vigorous growth in economic activity, particularly in manufacturing industry (up 5% a year). This should push up consumption of heavy fuel oil in industry, of motor fuels in the transport sector and of heating oil.

In contrast to other Community Member States, in Ireland the consumption pattern for oil products will shift towards heavier products after 1990 with the surge in demand in the industrial sector.

For the moment natural gas from the Kinsale offshore field has been ousting oil products at power stations or else been used primarily as a feedstock to produce ammonium and fertilizers. In the medium term it is likely to be used increasingly in the more traditional industrial, residential and tertiary markets.

As a result, with the coming on stream of the Moneypoint station in 1985 one can expect more coal to be burnt in power stations as a substitute first for oil products and then, in due course, for natural gas. At the same time the moves to promote conversion to coal in industry should increase coal's share of final energy consumption in the industrial sector.

Finally, peat production should continue at present-day levels, with one third of the output going to the domestic sector and the other two thirds to power stations.

48.   However, not even this combination of continued peat cutting and exploitation of Ireland's natural gas fields will be enough to reduce Ireland's dependence on imported energy supplies. On the contrary, increasing demand will probably push dependence up from 63% in 1983

to 72% in 2000. But exploratory drilling is continuing in the Irish section of the offshore fields. There is therefore still every chance that new oil or gas fields could be struck and further reduce Ireland's dependence on imported energy supplies in the long run.

Primary energy supplies in Ireland
(in million toe)

	1973	1980	1983	1990	2000
Domestic production	0.7	2.0	2.9	2.8	3.4
of which: peat	0.65	1.1	1.0	1.1	1.3
natural gas	-	0.7	1.8	1.4	1.7
Net imports	6.0	6.5	5.0	7.4	8.8
of which: coal	0.5	0.8	1.0	2.1	2.9
oil	5.5	5.7	4.0	5.3	5.9
Stock change	0.3	0.0	0.1	-	-
Total	7.0	8.5	8.0	10.2	12.2

Almost the entire anticipated extra demand for energy will probably have to be met by increasing imports, with the increase being shared equally between solid fuels and oil products.

(g) **Italy**

49. Gross energy consumption in Italy can be expected to increase by 1.8% a year on average between 1983 and 2000, assuming that GDP increases by 2.8% a year over the same period.

This would make the Italian economy roughly 17% less energy-intensive by the end of the century, with the primary energy ratio falling from 835 kgoe/10^6 Lit[24] in 1983 to 698 in 2000, following a 15% fall between 1973 and 1982.

Final electricity consumption is likely to increase by roughly 2.6% a year between now and 2000, with electricity's share of final energy consumption increasing to 17% by the end of the century compared with 14% in 1983.

Gross primary energy consumption in Italy (including bunkers)
(in million toe)

	1973	1980	1983	1990	2000
Solid fuels	8.1	11.0	12.8	21.5	31.0
Oil	102.2	97.0	86.3	83.7	81.6
Natural gas	14.2	22.7	22.4	32.7	35.1
Nuclear energy	0.9	0.7	1.6	5.4	19.4
Hydroelectric power and others	3.8	4.8	5.3	6.2	6.2
Total	129.2	136.2	128.4	149.5	173.3

----

[24] kgoe/10^6 Lit = kilogram of oil equivalent per million Italian lire at 1975 prices.

50.   Sharp increases in demand for coal and for natural gas will be the
main changes in the energy consumption pattern in Italy up to 1990.
After that   under the reference scenario solid fuels and nuclear
energy will cover virtually the whole increase in demand.   Little
change is likely for any of the other fuels.

Consumption of oil products in Italy is expected to fall by about 10%
over the whole period, with light and medium products gradually
increasing their share at the expense of heavy products, which will
suffer from the severe slump in consumption at power stations.

Power stations can be expected to make a massive shift away from oil
products and towards solid fuels. which will also continue their
penetration of industrial markets.

Natural gas too should increase its share of the industrial market,
where total energy demand is expected to rise by 1.7% a year
throughout the period from 1983 to 2000.   There should also be a
steadily expanding market for gas in the residential and tertiary
sector over the period to the end of the century.

Finally, natural gas is also expected to make inroads into power
stations for a limited period, pending the expansion of the
industrial and domestic markets.

51.   As the nuclear programme is put into action in the medium term and
with the exploitation of indigenous oil, gas. hydroelectricity and
geothermal resources, Italy should gradually reduce its dependence on
imported energy from 81% in 1983 to 76% in 2000.

Primary energy supplies in Italy
(in million toe)

	1973	1980	1983	1990	2000
Domestic production	19.3	17.5	19.1	25.9	41.2
of which: natural gas	12.6	10.2	10.6	10.0	10.0
nuclear energy	0.9	0.7	3.7	5.4	19.4
hydroelectric power and others	3.8	4.8	5.3	6.2	6.2
Net imports	112.1	118.9	104.8	123.6	132.1
of which: coal	7.7	10.8	12.1	20.0	29.5
oil	102.7	95.8	79.6	79.7	77.1
natural gas	1.6	11.8	12.1	22.7	25.1
Stock change	-2.2	-0.2	4.5	..	..
Total	129.2	136.2	128.4	149.5	173.3

Electricity - whether generated from nuclear. hydroelectric or
geothermal sources - will account for the entire increase in Italy's
energy output from indigenous sources.   Coal and natural gas imports
will probably at least double between 1983 and 2000, with oil imports
holding steady.

(h) **Luxembourg**

52. Assuming an average annual growth rate of 2.2% for the economy as a
     whole between 1983 and 2000, energy consumption in Luxembourg should
     rise by 1.6% a year over the period.

     This would bring the primary energy ratio of the Luxembourg economy
     down from 30 kgoe/10^3 Lfr[25] in 1983 to 27 giving a 10% improvement in
     energy efficiency on top of the substantial (40%) reduction between
     1973 and 1983 caused by the restructuring of the steel industry.

     Electricity consumption too is expected to increase by an average of
     2% a year until the turn of the century. Electricity's share of final
     energy consumption should rise from 11% in 1983 to 12% in 2000.

              Gross primary energy consumption in Luxembourg
                         (in million toe)

	1973	1980	1983	1990	2000
Solid fuels	2.45	1.84	1.27	1.55	1.65
Oil	1.65	1.10	1.00	1.05	1.15
Natural gas	0.22	0.42	0.26	0.40	0.45
Hydroelectric power					
and others	0.18	0.27	0.31	0.35	0.45
Total	4.50	3.63	2.84	3.35	3.70

53. Assuming that the steel industry gradually returns to 1980
     production levels or thereabouts, there should be a slight increase
     in consumption of solid fuels.

     The underlying reasons for the slight increase in oil product
     consumption will be increased demand for motor fuel in the transport
     sector and an upturn in demand in most branches of industry, apart
     from steel.

     The higher level of natural gas consumption ·· which should return to
     close to 1980 levels by the end of the century – will be due partly
     to the replacement of oil products by alternative fuels in the
     residential and tertiary sector and partly to industrial growth.

     New and renewable sources of energy such as hydroelectricity and
     biomass, could double their share by the end of the century, though
     the volumes produced will still be small.

54. Another important feature is that the forecast increase in
     electricity demand should push up electricity imports.

----
[25] kgoe/10^3 Lfr = kilogram of oil equivalent per thousand Luxembourg
     francs at 1975 prices.

Primary energy supplies in Luxembourg
(in million toe)

	1973	1980	1983	1990	2000
Domestic production[26]	0.01	0.02	0.03	0.05	0.10
Net imports	4.50	3.61	2.79	3.30	3.60
Stock change	-0.01	-	+0.02	..	..
Total	4.50	3.63	2.84	3.35	3.70

In 2000 all Luxembourg's energy requirements apart from those
covered by hydroelectric power (0.03 million toe a year) or by
biomass will be covered by imports, most of them from other
Community Member States.

(i) **Netherlands**

55.   Assuming an average annual growth rate of 2.5% for the Netherlands
economy from 1983 to 2000, gross domestic energy consumption can be
expected to grow by an average of 1.3% a year over the same period.

In addition, consumption of bunker fuel for seagoing
shipping — which already accounts for one quarter of all oil
products sold in the Netherlands - could soar by 50% by the end of
the century, thus returning to more or less the 1973 level.

Restructuring of the economy coupled with substantial energy savings
should reduce the primary energy ratio of the Netherlands economy
from 26 kgoe/10^2 Fl[27] in 1983 to 20.5 in 2000, a reduction of over
20% to follow the almost 25% drop between 1973 and 1982.

Final electricity consumption is expected to grow by an average of
1.4% a year between 1983 and 2000, markedly slower than in the other
Member States None the less, electricity will continue to cover
around 11% of final energy demand throughout the period thanks to
its relative competitiveness.

---

[26] Hydroelectric power, biomass and recovered products

[27] kgoe/10^2 Fl = kilogram of oil equivalent per 100 guilder at constant
1975 prices.

Gross primary energy consumption in the Netherlands
(including bunkers) (in million toe)

	1973	1980	1983	1990	2000
Solid fuels	3.2	4.1	5.1	10.4	10.5
Oil	41.0	38.4	29.9	34.3	37.9
Natural gas	28.5	30.4	29.2	28.0	28.0
Nuclear energy	0.3	1.1	0.9	0.9	4.4
Hydroelectric power and others	0.1	0.3	0.6	0.8	0.8
Total	72.9	74.3	65.7	74.4	81.6
of which bunkers	11.5	9.3	8.1	11.0	12.0

56.   Much of the appreciable increase in coal consumption in the medium
term will be the result of conversion of some of the Netherlands
power stations to coal. Steam-coal consumption in industry should
also rise in the medium term.

Natural gas consumption in the Netherlands is expected to level off
since the domestic market is saturated and the forecast increase in
sales to indust y will only partly offset the lower offtake by power
stations as a result of a gradual switch to coal.

Demand for oil products should gradually pick up in the medium and
long term as demand for motor fuel and for bunker fuel for shipping
increases along with sales to the petrochemicals industry. Sales of
medium and light oil products should increase and sales of the
heavier fractions decrease.

It is assumed in the reference scenario that three more nuclear
power plants will be brought on stream by 2000 in line with the
plans of the Dutch government.

57.   The net result of this surge in nuclear output by 2000 coupled with
rapidly growing coal imports and gradually declining natural gas
output is that Dutch dependence on imported energy supplies is
likely to rise from 10% in 1983 to 42% in 2000.

Primary energy supplies in the Netherlands
(in million toe)

	1973	1980	1983	1990	2000
Domestic production	56.8	69.6	59.4	54.8	47.3
of which: natural gas	53.8	66.7	55.3	49.7	39.1
nuclear energy	0.3	1.1	0.9	0.9	4.4
Net imports	16.3	5.4	4.6	19.6	34.3
of which solid fuels[28]	1.7	4.1	4.5	10.4	10.5
natural gas	-25.3	-36.2	-26.2	-21.7	-11.1
oil	40.2	37.6	25.9	30.6	34.9
Stock change	-0.2	-0.7	1.7	..	..
Total	72.9	74.3	65.7	74.4	81.6

----
[28] A minus sign (-) indicates net exports.

The key element in the projection. however, is the assumption that natural gas production from the Dutch fields will gradually decline over the next 15 years. Since domestic demand can be expected to remain stable, exports will have to be scaled down in volume terms, despite the plans to import natural gas from Norway.

All the coal consumed in the Netherlands will have to be imported, just like the bulk of the oil processed at Dutch refineries.

(j) United Kingdom

58. The reference projection for the United Kingdom assumes an average annual growth rate of 2.4% for the UK economy between 1983 and 2000, with gross energy consumption increasing by an average of 1.1% a year over the same period.

As a result the primary energy ratio of the UK economy should fall from 1.75 kgoe/£[29] in 1983 to 1.40 in 2000 - in other words. a 20% reduction in energy intensity, following the 17% or so reduction between 1973 and 1983.

Final electricity consumption is expected to increase by 1.8% a year by the end of the century, with electricity's share in final energy demand rising from 14.5% in 1983 to 16.5%in 2000.

Gross primary energy consumption in the United Kingdom
(including bunkers) (in million toe)

	1973	1980	1983	1990	2000
Solid fuels	79.2	70.0	65.4	72.0	70.4
Oil	113.5	81.8	74.3	76.6	81.2
Natural gas	25.1	39.9	42.4	47.1	50.7
Nuclear energy	8.9	10.4	13.5	20.5	34.8
Hydroelectric power and others	0.3	0.3	0.4	0.4	0.7
Total	227.0	202.3	196.0	216.6	237.8

59. The assumed increase in nuclear electricity supply is the most striking feature of the reference projection for the UK energy market for the rest of this century.

This would have a direct effect on solid fuels   with the market in power stations continuing to increase up to 1990, then levelling off and, in the end, even declining appreciably.

Natural gas consumption in the United Kingdom should also grow steadily, with a progressive increase in sales to the domestic and tertiary sectors and further consolidation of its share of the industrial market.

----

[29] kgoe/£ = kilogram of oil equivalent per pound at 1973 prices.

Consumption of oil products should also increase slightly partly because of the upturn in diesel sales to road users, in kerosene sales to aircraft and in naphtha sales to the petrochemicals industry. This would mean a shift towards lighter products, with the proportion of heavy products declining from 30% at present to 24% in 2000.

60. The scale and variety of the United Kingdom's indigenous energy resources, plus the plans to expand the UK's nuclear capacity, should enable the United Kingdom to remain virtually self-sufficient in energy for a long time while to come.

The United Kingdom's primary energy supply
(in million toe)

	1973	1980	1983	1990	2000
Domestic production	113.0	196.1	232.9	215.9	231.5
of which: oil	0.7	79.7	117.0	90.0	88.0
natural gas	24.4	30.9	32.8	35.0	38.0
nuclear[30]	8.9	10.4	13.5	20.5	34.8
Net imports	112.4	12.7	-34.8	0.7	6.3
of which: oil	112.6	2.0	-43.7	-13.4	-6.8
natural gas	0.7	9.0	9.6	12.1	12.7
Stock change	1.6	-6.5	-2.1	-	-
Total	227.0	202.3	196.0	216.6	237.8

Coal production is expected to maintain its present level until 2000, with new, competitive capacity gradually replacing high cost pits.

Oil and gas production, which should level out in the mid-1980s, is assumed to fall gradually thereafter. Clearly this tendency could be modified by the discovery of new fields in the North Sea. The United Kingdom would at all events remain a net exporter of oil at the end of the century. For natural gas, on the other hand, despite considerable domestic resources imports should remain steady at the level contracted for 1990.

VIII. A STUDY OF SOME ALTERNATIVES TO THE REFERENCE PROJECTION

61. **The reference projection described in the preceding pages should on no account be confused with a single forecast, but should be considered as a plausible and consistent scenario among other possibilities.**

---

[30] A minus sign (-) indicates net exports.

Obviously, the results obtained reflect the assumptions on which the projection is based. One key aim of the study, however, was to assess the sensitivity of the projections to changes in the key assumptions, so as to identify more clearly the nature of the uncertainties in the long term picture.

(a)   Variations in the Economic Framework

(i)   economic growth

62.   The reference projection assumes an average economic growth rate for the Community of 2.6% a year between now and 2000.

A lower rate of growth would have the effect of reducing the GFCF (gross fixed capital formation), thus slowing down the rate of capital replacement. This would delay the penetration of new and more energy-efficient technologies. On the other hand, the slower rate of growth of industrial production would result in a reduced total energy demand and lower electricity consumption. The combined effects of these two opposing tendencies would depend on the rate of growth chosen and its duration. For example, an annual growth of 1.8% up to the year 2000 would reduce final energy consumption by approximately 50 mtoe compared with the reference projection. The lower rate of growth of electricity consumption would chiefly affect solid fuel consumption.

A higher rate of economic growth, for example 3.5% a year, would enable the Community's energy efficiency to be improved by comparison with the reference projection but would also result in higher production of goods and services. This would lead, by 2000, to a gross energy consumption 60 mtoe higher than in the reference projection. Most of the new requirements would be covered by hydrocarbons: oil products (24 mtoe) and natural gas (16 mtoe); solid fuels (+ 7 mtoe) and nuclear energy (+ 12 mtoe) would also share in the increase in demand, though to a lesser extent.

(ii)   economic structure

63.   In the reference projection, a restructuring of the economy towards services, manufacturing industry and capital goods would have a favourable effect on the energy intensity of the Community economy.

If, however, it were assumed that energy-intensive industries were to maintain their present market shares up to the year 2000, the Community's consumption of energy per unit of GDP would fall more slowly than expected after 1990. Even if the technological advances **within** sectors introduced into the reference case were still assumed to occur, the steady growth in volume of output of the energy-intensive industries would lead to an increase in energy consumption in the year 2000 of the order of 60 mtoe above the reference scenario. Solid fuels would enjoy increased markets (+ 18 mtoe) because basic industries, such as steel and non-metallic ore are heavy coal consumers. Oil (+ 18 mtoe), natural gas (+ 13 mtoe) and nuclear energy (+ 13 mtoe) would meet the rest of the increase in demand.

EUR-10: 2000	Reference Projection	"Economic" alternatives	
		high GDP	unchanged GDP structure
Annual rate of increase of GDP	+2.6%	+3.5%	+2.6%
Gross energy consumption (mtoe)	1136	1195	1198
of which: solid fuels	264	271	282
oil	439	463	457
natural gas	196	212	209
nuclear	215	227	228
Net oil imports (mtoe)	330	354	348
Dependence on imported energy	45%	46%	46%

(b)   Variations in the Average Crude Oil Import Price:

64. The central assumption for the average crude oil import cost used in the "reference projection"[31] led to a $35/bbl level in 2000 (at 1983 prices).

Two alternative price levels were chosen: $20 and $50/bbl, by 2000.

(i)   low oil price

65. Assuming that crude oil prices were to fall from their 1983 level to $20/bbl in 1990 and to maintain that level in real terms until 2000, the Community's GDP would - all other things being equal - be higher than in the reference scenario because of reduced pressure on the balance of payments. This would lead to an additional net increase of about 80 mtoe in energy demand.   But because of the favorable oil price, oil demand would grow more rapidly than energy consumption as a whole notably at the expense of coal.

Oil consumption could thus increase by more than a quarter, reaching 540 mtoe and covering 44% of energy demand in 2000, instead of 39% as in the reference projection.   This increase in consumption could only be covered by imports and would thus result in an increase in the Community's energy dependence.

As regards the world energy balance, this scenario would have similar results:

- a drop in solid fuels demand, which would only increase by 8% between 1983 and 2000 instead of the 25% indicated in the reference projection.   This change would be due in particular to electricity being less competitive than oil or gas for satisfying local needs.   This, combined with the likelihood of slowdowns in the construction of coal-fired power-stations would result in a

----

[31] To arrive at final consumer prices, assumptions had to be made of course about transport and refinery costs, refinery netbacks, distribution costs and taxation. Given the uncertainties in all these areas it seemed reasonable to assume no major change in the present relationship between these costs and the basic price of crude.

lower consumption of coal in power stations. Solid fuels, especially coal, would also experience stiffer direct competition in the industrial heat markets;

- a fairly sharp recovery in demand for oil and (to a lesser extent) gas, which would cover both the increase in energy demand and reduced coal demand.

(ii) high oil price

66. If, on the other hand, oil prices were to grow much more rapidly than anticipated and reach $40/bbl in 1990 and $50/bbl in 2000, the real GDP growth rate for the Community would be reduced because of the inflationary effects and the additional balance-of-payments transfers.

In view of the cumulative effect of the reduction in GDP and the increase in the price of energy, the Community's gross energy consumption in 2000 would be approximately 70 mtoe lower than that estimated in the reference projection. With a crude oil price of $50 per barrel, the competitiveness of nuclear energy, natural gas and coal should be enhanced vis-à-vis competing oil products.

On the basis of the indexing systems adopted in the reference projection, natural gas would enjoy a very marked improvement in its threshold of competitivity. This could even result in sharp competition between gas producers and downward pressure on prices, which would increase its substitution for oil products for all non-specific uses.

The fact that electricity would be more competitive for heating would lead to an increase in its share in meeting final energy needs. This increase in production would be covered by both nuclear energy and solid fuels. The latter could also find themselves more competitive on industrial markets.

Under these circumstances oil consumption would be reduced by nearly a third, remaining limited to 300 mtoe. This would enable oil import dependence to be reduced to close to 40%.

EUR-10: 2000	Reference Projection	"Petrol price" alternatives	
		low price	high price
Average import price of crude oil	$35/bbl	$20/bbl	$50/bbl
Gross energy consumption (mtoe)	1136	1218	1066
of which: solid fuels	264	229	275
oil	439	539	306
natural gas	196	216	226*
nuclear	215	212	230
Net oil imports (mtoe)	330	439	194
Dependence on imported energy	45%	51%	39%

* This presupposes that natural gas prices are aligned downwards on marginal production costs (see Chapter VI, point (a)).

(c) Alternatives Regarding Natural Gas Prices

67. The system for determining natural gas prices is another factor which is open to conjecture. The reference projection assumes that the present system, based on links between natural gas prices and those of competing oil products, will continue.

This assumption leads to natural gas maintaining a share of approximately 18% in the Community energy market.

It is, however, possible that natural gas would increase its market share **if its relative price were reduced** and disconnected from oil prices.

Even in the context of the reference projection, it is possible to imagine that in case of a structural surplus in natural gas supply, its market penetration could be increased by some decoupling of gas prices from oil and aligning them with competing products in specific markets: coal, heavy fuel oil and gas oil for heating. Such an approach could reduce gas prices to consumers to drop by about 8% (compared with the reference case), increasing the demand for natural gas by 20 mtoe in 2000. Oil products and coal would find their markets reduced by 15 and 5 mtoe respectively. In this case the proportion of Community energy needs covered by gas would reach 20% in 2000.

Under the assumption that the oil price were to reach $50/bbl in 2000, all known potential sources of natural gas would become competitive. Increased competition between producers could lead to a more general "decoupling" of gas from oil prices, thus reducing its price by an estimated 29% (again compared with the reference scenario).

In this case, an additional 15 mtoe of oil could be substituted compared with the "high oil price" alternative (see section 66). A certain drop in coal consumption would also be possible.

The natural gas market could thus represent 24% of energy supply in
2000. This is , however, an extreme case which could only come about
in conjunction with considerable diversification of natural gas
supplies.

EUR-10: 2000	Reference Projection	consumer price alternatives	
		relative changes in natural gas prices	
	-	-8%	-29%
Gross energy consumption (mtoe)	1136	1136	1066
of which: solid fuels	264	259	270
oil	439	424	290
natural gas	196	216	256
nuclear	215	215	222
Net oil imports (mtoe)	330	315	178
Dependence on imported energy	45%	45%	40%

* Under the "high oil price" scenario.

(d)   Alternatives Regarding the Role of Solid Fuels

68.   An increase in coal consumption in the Community between 1983 and
2000 is an essential objective of energy policies as assumed in the
reference projection. If succcessful coal would not only replace oil
and gas in power stations but would also enjoy a certain revival in
industry. The reference scenario suggests that on this basis solid
fuels (coal, brown coal and peat) will maintain their present market
share of around 24%.

Lower oil prices could impact significantly on this outlook. Other
major questions relate to the impact of environmental measures and
possible future constraints on nuclear power penetration.

(i)   strict emission controls on large industrial plants

69.   The   Commission's   present   proposals   on   the   tightening-up   of
atmospheric emission controls on large industrial plants could
influence the coal and oil markets in the future.

There is a degree of uncertainty as to the way in which the Member
States would apply the proposed standards if they were adopted.

There are a number of options for industrial concerns and electricity
undertakings:

-    to replace the fuel used by an alternative, less polluting one;
natural gas could become attractive in this case;

- turning to fuels with low sulphur content, involving a cost increase of from 4 to 8%;

- installing desulphurization units, the investment cost of which can vary, depending on the size of installation, from $90 to $200/KWe.

A combination of these various measures would not lead to any appreciable change in gross energy consumption in 2000, but would result in changes in the balance between fuels compared with the reference projection.

In the electricity sector, stricter emission controls on conventional power stations will probably lead to increased production costs, varying according to country between 8 and 12%. In this case overall final electricity demand could drop by 3 to 4% by the year 2000 to the advantage of natural gas. As far as electricity production itself is concerned, there could be some substitution of nuclear energy for coal for base-load production, while marginal quantitites of coal and fuel oil could be replaced by natural gas for middle load.

In the same context, gas would also increase its market share in industry at the expense of coal and oil. The final result of this would be a level of coal consumption in 2000 close to 225 mtoe instead of the 265 mtoe arrived at in the reference projection.

Oil consumption would also drop by some 10 mtoe, in favour of natural gas and nuclear energy, which would each increase their respective market shares to much the same extent.

(ii) A slow-down in nuclear programmes

70. The use of solid fuels could, on the other hand, increase more rapidly if for one reason or another there were to be a slow-down in nuclear programmes. Assuming that major delays, postponements or even cancellations of part of the future programmes were to take place, this could result in an additional demand for coal to cover a growing share of base-load electricity production.

In this extreme case, consumption of solid fuels could approach 310 mtoe in 2000. Natural gas needs could also show a slight increase.

EUR-10 : 2000	Reference projection	Low coal	High coal
Modified assumptions	–	Tightening up of emission standards	Delays in nuclear programmes
Gross energy consumption (mtoe)	1136	1136	1133
of which: solid fuels	264	223	309
oil	439	430	444
natural gas	196	226	208
nuclear	215	235	150
Net oil imports (mtoe)	330	321	335
Dependence on imported energy	45%	44%	50%

(e) Analysis of Energy Efficiency

71. On the basis of the assumptions adopted in the reference projection, average (primary) energy efficiency in the Community should improve by about 20% by the turn of the century **assuming the maintenance of present energy saving policies.**

Although this improvement would be substantial it would not exhaust the scope for more efficient energy use in the different sectors of final energy demand. If more vigorous energy saving policies were to be introduced generally in the Community it would be possible to envisage savings of 100 mtoe or more compared with the reference projection. This would result from higher levels of insulation in buildings, heat recovery, district heating and combined heat and power, greater use of heat pumps and improvements in electronic control as well as organisational techniques.

72. The largest share of the savings could be made on the basis of technologies already available within the Community and which could be commercialised more widely. **The most important sector for savings would be the residential sector,** which would reduce directly the consumption of oil, gas and electricity. Solid fuels would be affected indirectly and to a less significant extent through an expected reduction in electricity consumption.

73. It should be underlined however that without the maintenance of present energy-saving policies – building regulations, support for development and commercialisation of technologies and so on – even the level of savings indicated in the Reference Scenario would not be realised. If this occurred, it would adversely affect primary energy demand, energy import dependence and energy prices.

The determining factor here will be the level of investment proposed and the resulting speed of capital replacement in the industrial, transport, services and domestic sectors.

IX. THE COMMUNITY'S PLACE IN THE WORLD

74. In 2000, the Community should be consuming 11% of the energy produced in the world and producing only 5.5%. In 1970, the corresponding figures were 16% and 6% respectively.

In spite of considerable improvements which have been made in its supply structure during the last ten years - a trend which is expected to continue during the 1990s - the Community is likely to remain dependent on the outside world (in other words the international energy market) for around half its supplies.

75. Between 1983 and 2000, the world economy is assumed to grow by 3.3% a year, entailing a growth of from 2.4% a year in **energy demand**, which will reach 10.8 thousand million toe in 2000.

Gross world consumption of primary energy
in million toe (reference projection)

	Solid fuel	Oil	Natural gas	Nuclear	Hydro & other	Total
Situation in 1983:						
Industrialized countries	949	1642	695	200	108	3594
Developing countries	109	535	155	7	447	1253
Centrally-planned economies	1171	629	486	33	106	2425
World	2229	2806	1336	240	661	7272
Projection for 2000:						
Industrialized countries	1530	1677	847	589	231	4874
Developing countries	238	919	329	38	709	2233
Centrally-planned economies	1683	840	887	147	156	3715
World	3453	3436	2063	774	1096	10822

Within the industrialised countries the level of economic growth is assumed to be close to 3% a year, with 2.6% in the European Community, 2.8% in the USA, over 3% in Japan and around 3% in other countries. Energy consumption in these countries together would grow by 1.8-2% a year, with a sharper "decoupling" of energy demand from GDP growth in the EC and in the USA than in the other countries.

Consumption of all forms of primary energy should expand in all parts of the world, with the (partial) exception of oil, the consumption of which will be flat in the industrialised world - under the reference case. But this is subject to major uncertainties with respect to economic growth, energy prices, energy production capacities, energy savings etc... World energy demand could therefore lie in a range of 10-12 thousand million toe.

If energy demand were to reach 12 thousand million toe, the additional demand for fuel, of more than 1 000 mtoe over and above the figures in the table above would probably be covered by an increase in oil and gas consumption. This is because coal demand in the industrialized and centrally planned economies in the reference projection is already expected to be pushing up against capacity constraints.

76. Trends in **world energy production,** linked to the reference demand scenario set out above, should not diverge greatly from past trends, because:

- the industrialized countries would only produce 75% of their energy needs at the turn of the century;

- developing countries would continue to account for nearly 60% of world oil production, of which about half would be intended for export;

- the centrally-planned economies would remain self-sufficient and would even be able, on balance, to remain net exporters of solid fuel and natural gas.

World production of primary energy, in million toe
(reference projection)

	Solid fuel	Oil	Natural gas	Nuclear	Hydro & other	Total
Situation in 1983:						
Industrialized countries	956	765	637	200	108	2670
Developing countries	91	1252	199	7	447	1996
Centrally-planned economies	1178	750	516	33	106	2583
World	2225	2767	1352	240	661	7245
Projection for 2000:						
Industrialized countries	1578	671	651	589	231	3720
Developing countries	158	1925	431	38	709	3261
Centrally-planned economies	1720	825	1005	147	156	3853
World	3456	3421	2087	774	1096	10835

* Differences between totals for consumption and production are due to rounding and statistical discrepancies.

Assuming that there were to be an additional energy demand of 1 000 mtoe to be satisfied by 2000, world oil production would have to rise to 4.000 mtoe and gas production to 2.500 mtoe.

From a technical point of view one third of the increase required in oil production could probably be covered by the industrialized countries and the other two thirds by the developing countries. For natural gas the increases in production would probably have to be shared equally between the industrialized and the developing countries. Whether these increases would be forthcoming in practice would depend on economic factors (notably price) and political conditions.

77. Within the framework of the assumptions made in the reference projection, our conclusion is that the world energy market could supply the volumes of energy necessary to meet the needs of the various parts of the world, including the Community. But the margins for manoeuvre would not be considerable. There is no guarantee that the investments required to maintain or increase oil or gas capacity in the developing countries will be made in the coming years, particularly in a lower price environment than assumed for the reference case.

Even a small percentage increase in world oil consumption - the result of a higher rate of economic growth or a lower consumption of other forms of energy than assumed in the reference scenario - could put severe pressure on the oil production capacity of the Middle East countries, which (owing to their large reserves and low production costs) remain the world's marginal source of oil. So there is still a risk of much tighter oil market conditions in the 1990s. Sustained low oil prices could bring this about in the absence of appropriate policy actions to reduce the risks.

ANNEX: Summary energy balances for the Community (EUR-10)

ENERGY SUPPLY/DEMAND BALANCE   EUROSTAT   1973

(Mio Toe)	SOLID FUEL	PETROL PRODCT	GAS	NUCLR ENERGY	HEAT	HYDRO+ OTHERS	RENEW ENERG	TOTAL
PRIM PRODUCT	197.7	13.1	112.2	17.7	–	9.1	1.2	351.0
TOT IMPORTS	42.6	732.6	29.3	–	–	2.9	–	807.4
TOT EXPORTS	23.6	136.3	25.3	–	–	2.2	–	187.4
STOCK CHANGE	5.3	-8.1	-0.4	–	–	–	–	-3.2
GROSS CONSUM	222.0	601.3	115.8	17.7	–	9.8	1.2	967.8
BUNKERS	–	37.4	–	–	–	–	–	37.4
INLAND CONSM	222.0	563.9	115.8	17.7	–	9.8	1.2	930.4
ELEC POWER S	101.3	75.0	30.6*	17.7	-3.2	-80.4	1.2	142.2
OTHER TRANSF	32.7	13.5	-33.7	–	–	–	–	12.5
ENERG INDUST	2.5	33.8	12.7	–	–	14.2	–	63.2
FINAL CONSUMP	85.4	439.3	107.4	–	3.2	76.0	–	711.3
– NON ENERGY	3.4	60.7	6.2	–	–	–	–	70.3
– INDUSTRY	45.6	104.7	57.9	–	0.9	38.6	–	247.7
– TRANSPORT	1.0	125.0	0.1	–	–	2.0	–	128.1
– HOUSEHOLD	35.4	148.9	43.2	–	2.3	35.4	–	265.2
STAT DIFFER	0.1	2.3	-1.2	–	–	–	–	1.2

* of which: gas derived from coal = 7.1 mtoe

ENERGY SUPPLY/DEMAND BALANCE   EUROSTAT   1980

(Mio Toe)	SOLID FUEL	PETROL PRODCT	GAS	NUCLR ENERGY	HEAT	HYDRO+ OTHERS	RENEW ENERG	TOTAL
PRIM PRODUCT	185.1	91.1	129.2	42.7	–	12.3	1.7	462.1
TOT IMPORTS	67.2	596.6	81.7	–	–	5.1	–	750.6
TOT EXPORTS	19.9	158.7	41.1	–	–	3.7	–	223.4
STOCK CHANGE	-9.7	-9.0	-0.5	–	–	–	–	-19.2
GROSS CONSUM	222.7	520.0	169.3	42.7	–	13.7	1.7	970.1
BUNKERS	–	26.2	–	–	–	–	–	26.2
INLAND CONSM	222.7	493.8	169.3	42.7	2.1	13.7	1.7	943.9
ELEC POWER S	130.1	60.9	31.3*	42.7	-3.9	-96.9	1.7	165.9
OTHER TRANSF	26.4	6.4	-25.0	–	–	–	–	7.8
ENERG INDUST	1.2	31.2	9.6	–	–	17.8	–	59.8
FINAL CONSUMP	65.0	395.3	153.4	–	3.9	92.8	–	710.4
– NON ENERGY	2.7	49.4	7.9	–	–	–	–	60.0
– INDUSTRY	38.4	77.6	66.8	–	1.4	42.8	–	227.0
– TRANSPORT	0.2	150.7	0.3	–	–	2.4	–	153.6
– HOUSEHOLD	21.8	117.4	76.5	–	2.5	47.6	–	265.8
STAT DIFFER	+1.9	+0.2	+1.9	–	–	–	–	+4.0

* of which: gas derived from coal = 6.5 mtoe

ANNEX: Summary energy balances for the Community (EUR-10)

ENERGY SUPPLY/DEMAND BALANCE   EUROSTAT   1983

(Mio Toe)	SOLID FUEL	PETROL PRODCT	GAS	NUCLR ENERGY	HEAT	HYDRO+ OTHERS	RENEW ENERG	TOTAL
PRIM PRODUCT	174.0	132.5	119.8	76.1	-	12.0	1.7	516.1
TOT IMPORTS	56.7	481.4	78.2	-	-	5.9	-	622.2
TOT EXPORTS	17.6	192.6	30.0	-	-	4.1	-	244.3
STOCK CHANGE	-0.9	+17.0	-2.8	-	-	-	-	13.1
GROSS CONSUM	212.2	438.3	165.2	76.1	-	13.8	1.7	907.3
BUNKERS	-	22.4	-	-	-	-	-	22.4
INLAND CONSM	212.2	415.9	165.2	76.1	-	13.8	1.7	885.0
ELEC POWER S	133.9	36.9	26.2*	76.1	-3.7	-99.1	1.7	172.0
OTHER TRANSF	21.7	3.6	-19.3	-	-	-	-	6.0
ENERG INDUST	0.8	26.3	9.9	-	-	18.3	-	55.3
FINAL CONSUMP	55.8	349.1	148.4	-	3.7	94.6	-	651.6
- NON ENERGY	1.3	47.8	8.3	-	-	-	-	57.4
- INDUSTRY	35.7	49.3	59.6	-	1.2	40.6	-	186.4
- TRANSPORT	0.1	152.8	0.3	-	-	2.4	-	155.6
- HOUSEHOLD	18.2	101.3	79.0	-	2.5	51.6	-	252.6
STAT DIFFER	+0.5	-2.1	+1.2	-	-	-	-	-0.4

* of which: gas derived from coal = 4.5 mtoe

ENERGY SUPPLY/DEMAND BALANCE   REFERENCE PROJECTION   1990

(Mio Toe)	SOLID FUEL	PETROL PRODCT	GAS	NUCLR ENERGY	HEAT	HYDRO+ OTHERS	RENEW ENERG	TOTAL
PRIM PRODUCT	175	111	115	145	-	13	4	563
TOT IMPORTS	79	455	103	-	-	5	-	642
TOT EXPORTS	12	125	28	-	-	6	-	171
STOCK CHANGE	-	-	-	-	-	-	-	-
GROSS CONSUM	242	441	190	145	-	12	4	1034
BUNKERS	-	28	-	-	-	-	-	28
INLAND CONSM	242	413	190	145	-	12	4	1006
ELEC POWER S	160	20	27	145	-10	-121	2	223
OTHER TRANSF	23	4	-16	-	-	-	-	11
ENERG INDUST	1	26	10	-	0	20	-	58
FINAL CONSUMP	58	363	169	-	10	113	2	714
- NON ENERGY	2	50	10	-	-	-	-	62
- INDUSTRY	40	57	66	-	4	48	-	216
- TRANSPORT	-	166	0	-	-	3	-	169
- HOUSEHOLD	16	90	92	-	6	61	2	267
STAT DIFFER								

ANNEX: Summary energy balances for the Community (EUR-10)

ENERGY SUPPLY/DEMAND BALANCE REFERENCE PROJECTION 1995

(Mio Toe)	SOLID FUEL	PETROL PRODCT	GAS	NUCLR ENERGY	HEAT	HYDRO+ OTHERS	RENEW ENERG	TOTAL
PRIM PRODUCT	173	111	111	178	−	14	5	591
TOT IMPORTS	94	454	104	−	−	5	−	657
TOT EXPORTS	12	124	21	−	−	5	−	161
STOCK CHANGE	−	−	−	−	−	−	−	−
GROSS CONSUM	255	441	194	178	−	14	5	1086
BUNKERS	−	28	−	−	−	−	−	28
INLAND CONSM	255	413	194	178	−	14	5	1058
ELEC POWER S	171	19	22	178	−13	−134	3	245
OTHER TRANSF	22	4	−15	−	−	−	−	11
ENERG INDUST	2	28	10	−	−	22	−	62
FINAL CONSUMP	60	362	177	−	13	126	2	740
− NON ENERGY	2	54	11	−	−	−	−	68
− INDUSTRY	43	55	69	−	5	53	−	225
− TRANSPORT	−	174	0	−	−	3	−	177
− HOUSEHOLD	15	79	97	−	8	70	2	270
STAT DIFFER	−	−	−	−	−	−	−	

ENERGY SUPPLY/DEMAND BALANCE REFERENCE PROJECTION 2000

(Mio Toe)	SOLID FUEL	PETROL PRODCT	GAS	NUCLR ENERGY	HEAT	HYDRO+ OTHERS	RENEW ENERG	TOTAL
PRIM PRODUCT	172	108	108	215	−	14	7	625
TOT IMPORTS	104	454	104	−	−	4	−	667
TOT EXPORTS	12	124	16	−	−	4	−	156
STOCK CHANGE	−	−	−	−	−	−	−	−
GROSS CONSUM	264	439	196	215	−	14	7	1136
BUNKERS	−	29	−	−	−	−	−	29
INLAND CONSM	264	410	196	215	−	14	7	1107
ELEC POWER S	178	18	15	215	−15	−148	4	266
OTHER TRANSF	23	4	−15	−	−	−		13
ENERG INDUST	2	29	11	−	−	24	−	65
FINAL CONSUMP	61	359	185	−	15	138	3	763
− NON ENERGY	2	58	12	−	−	−	−	72
− INDUSTRY	46	51	73	−	5	58	−	234
− TRANSPORT	−	178	0	−	−	4	−	182
− HOUSEHOLD	13	72	100	−	10	77	3	274
STAT DIFFER								

Summarized Energy Balance - EUROPEAN COMMUNITY

in million toe	1973[a]	1980[a]	1983[a]	1990[b]	1995[b]	2000[b]
I. Gross Energy Consumption	968.04	970.04	909.08	1034.0	1086.3	1135.7
– Bunkers	37.36	26.21	21.98	27.9	28.4	28.9
– Inland consumption	930.68	943.83	887.10	1006.1	1057.9	1106.8
II. Inland Energy Consumption	930.68	943.83	887.10	1006.1	1057.9	1106.8
– Solid fuels	221.97	222.68	211.99	242.0	254.8	264.5
– Oil	563.93	493.81	416.30	413.5	412.6	409.9
– Gas	115.83	169.26	165.34	189.6	193.9	195.9
– Primary electricity, etc.	28.95	58.08	93.47	161.0	196.6	236.5
III.Indigenous Production[1]	351.29	462.10	518.10	563.5	590.8	624.8
– Hard coal	171.16	153.31	143.06	139.5	136.2	134.6
– Lignite & peat	26.49	31.81	30.99	36.0	36.5	37.7
– Oil	13.17	91.10	132.51	111.5	110.8	108.0
– Natural gas	112.20	129.16	119.94	115.3	110.9	108.2
– Nuclear energy	17.73	42.67	76.06	145.0	177.6	215.1
– Hydro & geothermal[2]	9.38	12.39	13.87	12.6	13.7	14.2
– Others & renewables	1.16	1.66	1.67	3.6	5.1	7.0
IV. Net Imports[3]	619.91	527.15	377.90	470.5	495.5	510.9
– Solid fuels	19.00	47.28	39.80	66.5	82.1	92.2
– Oil	596.21	437.95	288.78	329.9	330.2	330.8
– Natural gas[2]	4.01	40.56	48.17	74.3	83.0	87.7
– Electricity[2]	0.69	1.36	1.87	-0.2	0.2	0.2
V. Stock changes[4]	3.15	19.21	-13.08	–	–	–
– Solid fuels	-5.32	9.72	1.14	–	–	–
– Oil	8.09	9.03	-16.99	–	–	–
– Gas	0.38	0.46	2.77	–	–	–
VI. Electricity Generation Input	235.13	280.72	288.60	366.5	405.8	443.5
– Solid fuels[5]	108.33	138.27	138.38	164.6	175.0	182.3
– Oil	75.04	60.91	36.86	20.4	18.7	17.8
– Natural gas	23.51	24.82	21.76	21.8	17.9	10.6
– Nuclear energy	17.71	42.67	76.06	145.0	177.6	215.1
– Hydro & geothermal[2]	9.38	12.39	13.87	12.6	13.7	14.2
– Others & renewables	1.16	1.66	1.67	2.1	2.9	3.5

Main indicators (related to long term objectives)

	1973–1963	1980–1975	1983–1980	1990–1983	2000–1990
Inland Energy annual growth rates	+4.6%	+1.9%	-2.1%	+1.8%	+1.0%
GDP annual growth rates	+4.7%	+2.9%	+0.4%	+2.4%	+2.8%
Energy-GDP ratio	0.98	0.65	–	0.75	0.37

	1973	1980	1983	1990	1995	2000
Share of oil in gross energy consumption	62.1%	53.6%	48.2%	42.7%	40.6%	38.6%
Share of coal and nuclear in in electricity production	53.6%	64.5%	74.3%	84.5%	86.9%	89.6%
Supply dependance on imports	64 %	54.3%	41.6%	45.5%	45.6%	45.0%

Sources: a. Statistical Office of the European Communities
         b. Energy 2000
Notes: 1. Production of primary sources, including recovered products.
         2. The conversion of electricity, including hydro and geothermal, is based on its
            actual energy content: 3600 kjoules/kWh or 860 kcal/kWh
         3. The (–) sign means net exports
         4. The (–) sign means a stock decrease
         5. Including coke oven gas and blast furnace gas (derived from coal)

CHAPTER I

REFERENCE PROJECTION: Key assumptions

A. GENERAL ECONOMIC FRAMEWORK

I. Introduction

1. The economic environment for the energy reference projection to 1990 was determined from the 1983 economic projections of the European Commission's Directorate-General for Economic and Financial Affairs. These projections, essentially of macro-economic aggregates, were established with the help of the COMET IV model and national economic budgets of 1983-84.

These projections were supplemented by a sectoral breakdown of the macroeconomic aggregates and by a scenario for the macro-economic environment 1990-2000.

II. The Economy

i) The international environment

2. The European economic outlook has to be set in an international context determined mainly by the price of oil, trade outside Europe and ECU/US dollar parities.

Price of oil
The price of imported crude oil is assumed to pass through three successive stages: fall in nominal terms until the mid-80s, followed by stability in real terms until 1990, and finally progressive growth in real terms to reach $35/bbl at 1983 prices in 2000.

International trade in the rest of the world
International trade, defined in terms of imports of goods outside the EEC area, is projected here to grow by 1990 at an average annual rate ranging from 3.8% in the OECD area (excluding the EEC, Japan and the United States) to 4.4% for Japan and 6.2% for the United States. After 1990, the same rates of growth broadly apply.

Economic activity in the rest of the world in the latter half of the 1980s is expected to be marked by an upturn in economic growth in the United States and Japan, at 2.7% and 4.7% on average respectively (1984-1990).

Dollar exchange rates

Estimating exchange rate trends is always a delicate matter, because it has to take into account elements such as inflation differentials and interest rates which are difficult to forecast in the long or even medium term. For COMET's requirements, however, assumptions had to be formulated for the period 1983-90: the ECU appreciating slightly against the dollar (by one tenth of a point per annum on average) because of a strengthening of the German mark, the guilder, the pound and, to a lesser extent, the Belgian franc.

ii) The European economic outlook

3.   As far as **internal demand** is concerned, over the whole period, and more particularly from 1984 to 1990, most Community countries are assumed to give preference to GFCF (gross fixed capital formation) rather than private consumption.

For the medium term, the model paints the following picture:

Trend of private consumption and GFCF

Average annual growth rate in % (83 - 90)	Private consumption	Gross fixed capital formation
EUR-10	1.9	3.2
FR Germany	2.0	3.6
France	1.9	1.9
Italy	2.7	2.7
The Netherlands	0.3	2.2
Belgium	0.9	2.7
United Kingdom	2.0	4.5
Ireland	0.7	3.7
Denmark	1.4	5.3
Greece	1.8	3.9

With the exception of Ireland, the results are very closely bunched, the maximum spread in GDP being only 0.8%.

4.   In the reference projection, **the Community external balance of goods and services** is influenced up to 1990 by two opposing movements:

- a slight deterioration in the terms of trade (-0.1% per annum);

- exports rising faster in volume terms than imports (+0.6% per annum).

The combined effect is an improvement in the Community balance of goods and services, which increases from + 0.6% of GDP in 1983 to 1.6% of GDP in 1990.   Over the period 1985-90, the trade balance of all countries except Italy and Greece moves back into surplus (at least in certain years).

Foreign trade

Average annual growth in %, 1983-90	Exports in volume terms	Imports in volume terms
Eur-10	3.8	3.1
FR Germany	4.3	3.4
France	3.3	2.0
Italy	5.2	5.2
Netherlands	3.5	1.6
Belgium	3.8	2.4
United Kingdom	2.7	3.7
Ireland	7.9	4.0
Denmark	3.3	2.4
Greece	2.4	2.8

5. Projected growth of **gross domestic product** is set out in the following table.

GDP growth rates

Average annual growth rate in %	1970-1980	1980-1983	1983-1990	1990-1995	1995-2000
FR Germany	2.0	0.0	2.6	2.8	2.8
France	3.6	0.8	2.1	2.8	2.9
Italy	3.2	-0.5	2.8	2.7	2.8
Netherlands	2.9	-0.6	2.0	2.8	2.9
Belgium	3.3	0.0	1.8	2.2	2.3
Luxembourg	3.2	-1.4	2.0	2.2	2.3
United Kingdom	2.0	1.4	2.2	2.5	2.5
Ireland	4.1	1.1	3.7	3.4	3.4
Denmark	2.3	1.8	2.0	2.5	2.7
Greece	4.8	-0.1	2.4	2.7	2.8
EUR-10	3.0	0.4	2.4	2.7	2.8

Historically, **the annual growth rate of Community GDP** fell from +4.7% on average in the years 1960-73 to +1.7% over the period 1974-83.

The reference projection, however, is based on an average growth rate of **+2.6%** per year over the period 1983-2000 (+2.4% per year from 1984 to 1990 and +2.8% per year from 1991 to 2000).

iii) <u>Sectoral trends</u>

6. The assumption of strong competition worldwide is a feature of the international environment. It points to increasing specialization of individual economies in capital goods and certain basic goods such as chemicals or non-ferrous metals, whose competitive production requires high levels of technology. In order to remain competitive, companies are assumed to adapt the structure of their fixed assets to the changing environment, which will create a more favourable climate for the expansion of investment goods.

   The growth of industrial exports and the general upturn in industrial production should encourage trade and expand the demand for transport and engineering services.

   Private consumption should be boosted by the general economic revival encouraging an accelerated demand for services.

   Similarly, building and civil engineering should benefit from the recovery in private consumption as well as from the expansion of business investment.

7. The assumptions for the future structure of GDP suggest an increase in the relative share of services, from 55% in 1980 to 59% in 2000, with a corresponding decline in the shares of agriculture (by 1%), building and construction (by 1%) and industry (by 2%).

   It is assumed that the **industrial restructuring** that started at the end of the 1970s will continue and that by 2000 the share of intermediate goods industries and primary processing (which are energy-intensive) will fall still further and the share of the capital goods industries will increase correspondingly.

   The share of the consumer goods industries will also be reduced.

GDP components and uses

EUR-10	1970	1980	1983	1990	2000
GDP components (%)					
- Agriculture	4.8	4.3	4.3	4.1	3.4
- Energy-intensive industries[1]	10.2	9.2	9.0	8.6	7.7
- Other industries	26.0	25.3	25.3	25.2	24.9
- Building and public works	8.3	6.3	5.9	5.6	5.4
- Services	50.7	54.9	55.5	56.5	58.6

B. FUTURE ENERGY PRICES

I.   General Considerations

1. It is possible to envisage a number of long-term paths for energy prices, depending on assumptions about the behaviour of the various factors determining equilibrium conditions on the energy markets, specifically:

   - the outlook for **energy supplies** which will depend on the level of marginal production costs, the speed of exploration and the success rate, the trend of reserves/production ratios, and so on;

   - the outlook for **demand** (which will be affected by economic growth and such factors as the emergence and penetration of new applications technologies); and

   - **government action** affecting supply and demand (regulatory requirements, incentive schemes, subsidies, aid, taxes, tariff policies, etc.).

2. To simplify the choices for the definition of the reference scenario it was agreed that the "base" price projections should not divert radically from recent trends[*].

   Three fundamental assumptions were also adopted for the long term:

   - the continuance of an international energy environment in which supply and demand operate as freely as at present;

   - the absence of any political event which might lead to increased intervention by governments or to crisis actions;

----
[1] Metals, non-metallic minerals, chemicals, pulp and paper.
[*] The study was completed before the sharp oil price falls of end-1985 and early 1986.

- the maintenance of current energy pricing and taxation policies in the Community.

3. The world-wide analysis carried out in this context[2] suggests that by the year 2000 hydrocarbons will still be the dominant energy source, meeting between 50 and 55% of energy requirements, with oil alone supplying 30 to 35%. Given this and the fact that in the majority of natural gas offtake contracts, price clauses are indexed to crude oil and/or certain refined oil products, the reference projection assumes that the **price of oil will remain the guide price** on the international energy market between now and the year 2000.

4. For each fossil fuel (oil, natural gas and coal) as well as for electricity, the set of **central** price-trend assumptions in the reference projection and the **alternative** price assumptions studied as variants are described in the text which follows.

II. Crude Oil and Oil Products

a) Cost of imported crude

5. "Scenarios to the year 2000" - the scenario study carried out in 1983 by the Commission's departments[3] - considered three different paths for the price of crude oil imported into the Community, which resulted in a price in 2000 between $33.3 and $36.6 per barrel at 1983 prices.

As there had been no substantial market changes since the assumptions underlying "Scenarios to the year 2000" were selected, it seemed appropriate to maintain a cif price for oil imported into the Community in 2000 of $35 per barrel at 1983 prices as the central assumption in the reference projection.

This end-of-the-century price level is the result of the following chain of developments:

- a fall in nominal prices from 1983 to 1985;

- stability in real prices from 1985 to 1990, and

- thereafter a steady increase in the real price of 2.5% per annum,[4]

----

[2] See Chapter V, The Community's Place in the World Energy Markets.
[3] See the Commission staff paper "Scenarios to the Year 2000 - a prospective study of the energy market in the European Community", 1983.
[4] This trend broadly corresponds to the central assumption made by OPEC's long-term Strategy Committee, by Chase Manhattan Bank in its study published in 1983 and by many energy companies. Even in 1985 (when this study was being prepared for publication) many of the companies still maintained this assumption for the longer-term despite the downward

assuming a moderate but sustained recovery in world demand for oil.

The results of these assumptions are presented in the following table:

Central assumption for the price path for crude oil imported into the Community ($US at 1983 prices, cif)

	1970	1975	1980	1981	1983	1985	1990	1995	2000
$US/bbl	3.5	17.7	34.7	38.9	30.1	26.5	27.3	30.9	35

To simplify matters, this central assumption for the Community was used for each Member State, although historically their average cif supply costs have been somewhat dispersed around the Community average.

6. The assumed trend in crude oil prices presupposes that:

- world demand for oil, which has been steadily falling during the last few years, will tend to stabilize, then grow again notably as the developing countries' requirements increase;

- the current OPEC surplus production capacity will have been managed by these countries in the interest of both oil-consuming and producing countries;

- there will be no spctacular unforeseen expansion in non-OPEC oil production in the next 15 years;

- there is no major political or economic crisis.

Some of these assumptions may be questioned. One key issue is how far the changes in the level of energy demand in the industrialized countries to date have been structural or conjunctural. If they are not as structural as is supposed, oil consumption could grow more rapidly, generating upward pressure on international prices.

On the other hand, it may be argued that oil demand will be concentrated increasingly on the specific markets of fuels and petrochemical feedstocks. And the developing countries may not increase their demands for oil so rapidly (either because of financial constraints or because of avoidance of oil-based economic growth). Such developments would lead to further contractions in oil requirements and would provoke a downward pressure on crude oil prices.

----

pressure on oil prices.

7. These points are made here simply to illustrate some of the
    uncertainty surrounding the central assumption about the long-term
    price of imported crude oil.

    Two **alternative price scenarios** bracketing the reference projection
    were therefore studied as variants on the central case.

    The **first variant** assumes fast-rising prices - by 10% per annum
    between 1986 and 1990 - in response to a relative reduction in supply
    or a rapid increase in world oil consumption.

    The **second variant** assumes that the price of oil will be further
    eroded in nominal terms up to 1990 as a result of a structural supply
    surplus. The price then remains stable in real terms.

    These alternative paths for the price of imported crude oil are
    summarized in the table below:

Two alternative paths for the cif price of crude oil imported into
the Community (US$ at 1983 prices)

US$/bbl	1975	1980	1981	1983	1985	1990	1995	2000
High Scenario	17.7	34.7	38.9	30.1	26.5	40	47	50
Low Scenario						20	20	20

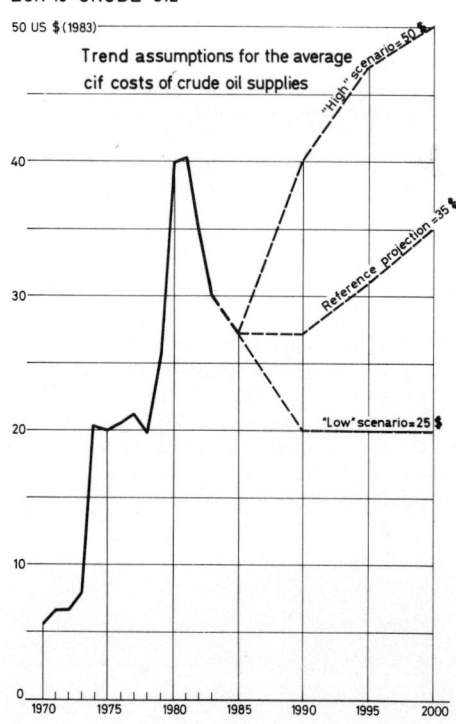

EUR-10: CRUDE OIL

Trend assumptions for the average
cif costs of crude oil supplies

Both these alternative
assumptions for the price
of imported crude have
obvious implications for
competing energy sources.
It can be reckoned that at
$20/bbl the current
sources of natural gas
supply will remain comp-
etitive but that some
investment for additional
production may be
abandoned. At $50/bbl
there would be not only
fierce competition between
gas, coal and oil, but
probably also competition
between gases from
different sources.

b) Consumer prices

8. **Consumer prices of oil products** are a function of transport and refining costs, gross ex-refinery margins, distribution costs and the tax component as well as the price of crude.

Given the many uncertainties about each of these elements, it seemed reasonable to suppose that present costs (excluding the price of crude itself) would be simply maintained in real terms. A priori this assumption may seem questionable with regard to the cost of refining: but it could be considered that the additional cost resulting from investments in plant conversion would be offset over time by a fall in operating costs. The implicit improvement in the rate of refinery capacity utilization, obtained in particular by closing surplus units, should have a favourable impact on operating costs.

The most difficult problem was how to define a conversion formula (linking the in-refinery cost of the crude oil processed (margin included) and the ex-refinery prices (before tax) of the principal petroleum products) which would show reasonable **ex-refinery margins on the processed crude.**

Community gross margins in 1983, expressed as the ratio of the sales income on the principal petroleum products to the average cost price per tonne processed were as follows:

Gross margins: Community average in 1983

	Petrol	Gas oil	Heavy fuel oil	Other products
EUR-10	1.19	1.01	0.63	1.12

Although the structure of the gross margin per tonne processed was found to be similar in all countries, the national values calculated are somewhat dispersed around the average ratios obtained for the Community, depending on local oil market conditions.

Gross margins: Community average and national ranges

	1978		1983	
	Average	Range	Average	Range
EUR 10				
Petrol	1.45	1.28 - 1.86	1.19	1.14 - 1.22
Gas oil & heat. oil	1.01	0.95 - 1.10	1.01	0.93 - 1.07
Residual fuel oil	0.60	0.57 - 0.66	0.63	0.61 - 0.65 [5]
Other products	1.22	1.00 - 1.59	1.12	1.06 - 1.18

----

[5] For one country, where the quantities were relatively small in Community terms, a ratio of 1.59 was obtained.

As the above table shows, the spread of national margins for the
various products narrowed considerably over five years, a fact to be
seen in conjunction with the significant increase in the share of
light products (petrol + kerosene + naphtha) in Community consumption
from 1978 to 1983.   Given this narrowing of the spread it was
considered reasonable to use for the longer-term a single set of
gross margins for the various products before tax for all Member
States. These were fixed as follows:

Gross margins: Community average, 1984–2000

	Petrol	Gas oil	Heavy fuel oil	Other products
EUR–10	1.20	1.00	0.6	1.12

EUR-10 : PETROLEUM PRODUCTS

Trend assumptions for consumer prices

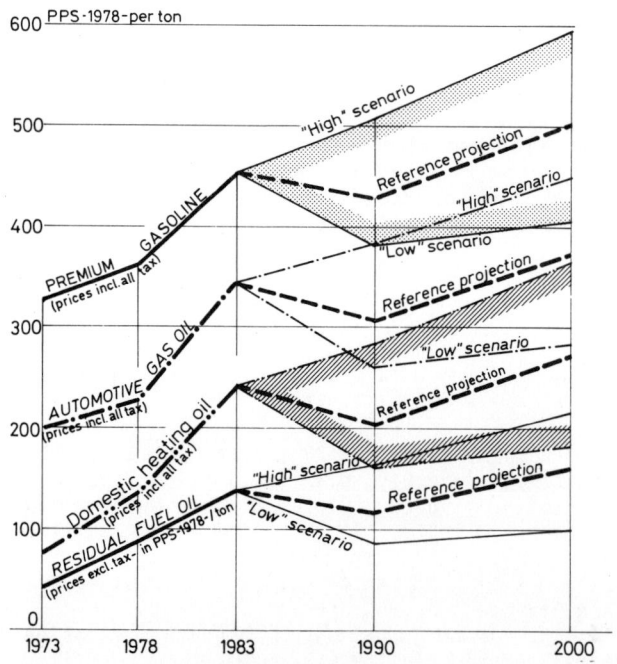

*Excl. tax, PPS (1978) per tonne.

III. Natural gas

a) Natural gas supply prices

9. The reference projection assumes that the price of oil will remain
the "guide price" for energy and that the formulas for **indexing gas
prices to crude oil** and/or certain oil products should continue to
apply until 2000.

The average cif procurement price paid by the gas industry in the
Member States varies from country to country in response to different
factors such as:

- the proportion of a country's gas supply accounted for by domestic
  production;

- the date when current supply contracts were concluded and the
  offtake and indexing clauses agreed;

- the breakdown of total gas imports by origin.

The various combinations of these factors which are dictated by each
country's own situation explain why the average gas price moves
differently from one country to another.

For the Community as a whole, the weighted average purchase price
fell slightly between 1980 and 1983.   In some countries, however,
changes in the structure of supply (Belgium and France) or a higher
internal offtake price (United Kingdom) led to an increase in the
weighted average price, which in 1983 erased the fall induced by the
formulas index-linking the price of gas to that of oil.

10. In view of the amount of uncertainty attaching to **future price
trends,** a number of simplifications were made in addition to linking
the price of oil to that of natural gas.

We supposed that in a medium-/long-term projection, the lags inherent
in the indexing formulas could be left out of account, and that in
the medium term, the indexing formulas adopted by each gas exporter
would be identical for all purchasers, irrespective of when the
contracts were signed.

Finally, simplified mechanical effects were allowed in order to
translate the impact of changes in the price of oil.   The simplified
mechanisms selected were as follows:

- for national internal production: the selling price of natural gas
  varies in proportion to that of heavy fuel oil and, as a result of
  further simplification, to that of crude oil;

- for imports from the Netherlands, Norway and the USSR: the changes would be proportional to those for two reference petroleum products: residual fuel oil and gas oil;

- for imports from Algeria: the absolute changes in the price of crude, expressed in dollars, would be passed on unaltered in the natural gas procurement price.

The table below summarizes the changes in the price of natural gas, compared with 1983, that should be taken into account in the assumptions selected for the reference projection.

Differences compared with 1983 in the unit procurement price of
natural gas in the Community, by origin
Reference projection
(in real terms based on 1983 prices)

	1990	1995	2000
Internal production	-9.3%	+2.7%	+16.3%
Imports from:			
- Netherlands)			
- Norway)	-8.1%	+2.4%	+14.3%
- USSR)			
- N. Africa	-0.5$/MBTU	+0.14$/MBTU	+0.87$/MBTU

By applying these various changes, country by country, and by weighting these various prices with the quantities of gas assumed from each place of origin[6], it was possible to establish an **average procurement price for natural gas,** in the Community, for the reference projection. This approach using a weighted average suggested that the average procurement price of natural gas would rise faster from 1983 to 1990 than the crude oil price as a result of the increased share of imports from North Africa, largely in the form of LNG which commands the highest prices.

Assumed trend of average procurement prices of natural gas in the
Community
($US/MBTU (gcv) at 1983 prices)

$/MBTU EUR-10	1980	1983	1990	1995	2000
	3.51	3.48	3.39	3.88	4.34

11.   In **the "low" alternative oil price scenario** ($20/bbl), the period 1985 to 1990 should see a major reduction in the average procurement price of gas in the Community. Given the structure of the indexing

---

[6] See COM(84)120 of 13 April 1984: Communication from the Commission to the Council concerning natural gas.

formulas, the cost advantage compared to oil tends to decline, especially for the most expensive places of origin. Thus, from 1990, liquified natural gas is no longer competitive in Belgium or France and may therefore disappear from the market. The delivery prices to Italy via the transmediterranean pipeline may, however, remain competitive.

In the **"high" alternative oil price scenario** ($50/bbl),the indexing formulas strengthen the price advantage of gas against oil. Moreover, the levels reached make it possible to consider exploiting Troll and other fields in the North Sea above the 62° parallel. Given the competitiveness between gas from different origins which results after 1990, all the average purchase prices of natural gas were stabilized at the producer's marginal operating cost, i.e. $5.9/MBTU.

Two alternative scenarios for the trend in the procurement prices of natural gas in the Community

($US at 1983 prices)

$US/MBTU	1980	1983	1990	1995	2000
High scenario	3.51	3.48	4.75	5.50	5.90
Low scenario			2.55	2.55	2.55

b)   Prices of natural gas delivered to consumers

12.   Three factors are decisive for setting **consumer gas prices:**

- the breakdown of each country's gas supply by place of origin;

- the structure of the national consumer market;

- general government economic policy and pricing and taxation policies.

An initial estimate of the quantities of gas coming from each place of origin was the basis for determining the average procurement price. To this have to be added, for each country, the transport, distribution and storage costs and the distribution companies' margin in order to obtain the average cost price of natural gas.

13.   The structure peculiar to each national market and the characteristics of the local gas industry lead not only to different cost prices for natural gas, but also to different pricing arrangements for the different consumer sectors, which reflect national circumstances. The price of natural gas is thus a direct function of the price of competing fuels on each particular market, i.e. residual fuel oil (often with a low sulphur content) in industry, gas oil in the heating sector and for certain industrial applications, electricity for small domestic users and steam coal in power stations.

Weighted average sales income, exclusive of tax, by consumption
sector for the whole Community, in $/GJ (gcv) at 1983 prices

	Industry	Households	Power stations	Weighted average sales income
1980	4.85	7.15	4.12	5.85

The structure of **gross margins on natural gas** sold to different
consumer sectors is fairly clear and broadly similar from one country
to another. In addition, they diverge reasonably little from the
Community average. Accordingly, and given the market trends that can
be predicted in the longer term, the same average assumption for the
breakdown of sales income by sector was used for all countries up to
the year 2000.

Index of average gross margins for natural gas, by consumption
sector, in the Community, at 1983 prices

	Weighted average sales income	Households	Industry	Power stations
All Community countries	100	125	80	70

This central assumption was modified, by sector, each time the gas
prices were likely in a given sector to be the same as the equivalent
price of the directly competing reference fuel. This was done in the
"low" oil price scenario.

To carry out this adjustment, the average sales income was reduced
(though not below a limit representing average procurement price plus
irreducible costs, i.e. inland transport, distribution and storage),
but the index for the gross margins in the different sectors was
assumed to be unchanged.

Finally, the price paid by the consumer was determined by adding in
the tax component (primarily VAT, except in Italy and Denmark).

EUR-10: NATURAL GAS Trend assumptions for consumer prices

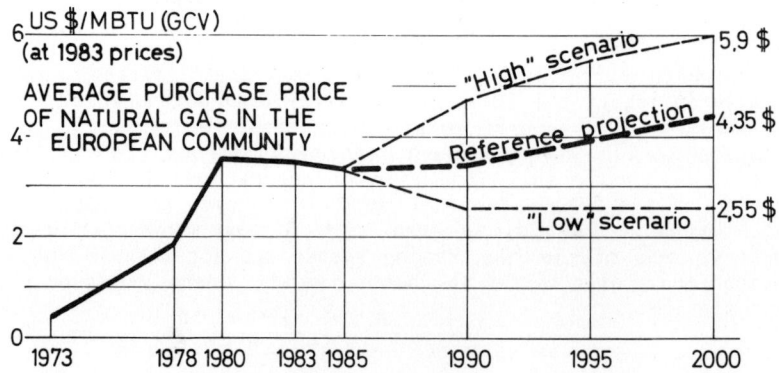

EUR-10: NATURAL GAS Trend assumptions for consumer prices

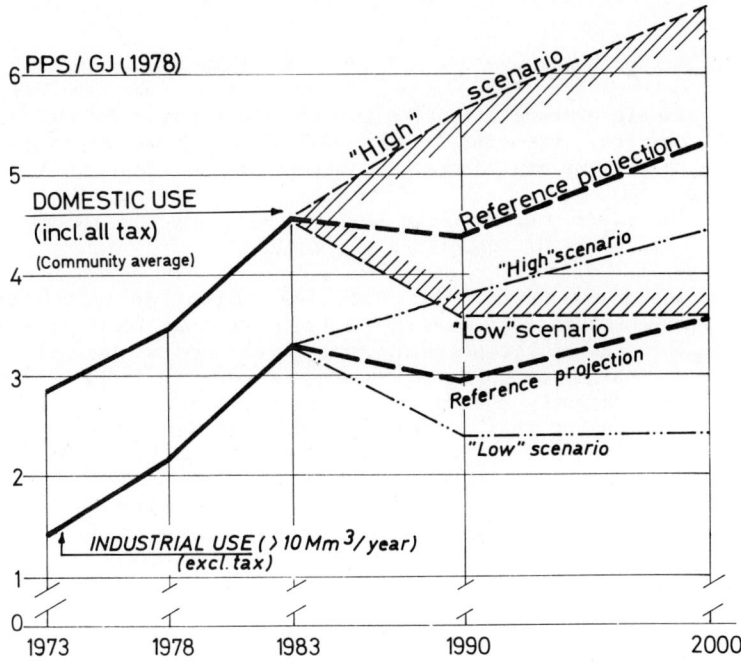

IV. Solid fuels

a) The cost price of coal

14. The **coal market** in fact comprises three distinct markets: coking
coal, house coal and steam coal.

The first two markets consist of specific users having very precise
requirements as regards the characteristics and quality of the
product consumed. For this reason, the prices of these types of coals
have always been higher than those of steam coal.

In addition, the coking coal market, which depends on the iron and
steel industry, and the house coal market are shrinking year by year
and will continue to decline in future.

We have therefore, to simplify matters, formulated assumptions
regarding the trend for steam coal only, coking coal and house coal
prices enjoying an advantage compared with the trend of the steam
coal price.

Steam coal from non-member countries has a cost advantage over
Community coal. The situation is expected to continue, and the price
of imported coal will thus remain the "guide price" for the coal
market.

Community coal will continue to carry higher costs and its price will therefore still have to be aligned in part to world prices. However, in view of ongoing changes in the UK mining industry, the difference between the British and the world price may tend to decrease after 1990.

Although the average cif price of imported coals has varied somewhat depending on the country of destination, we have established only **a single average cif price for the Community.** In addition, to simplify matters, the cost of maritime transport was taken to be constant in real terms and identical whatever the destination.

15. There are two possible ways of defining the long-term level of the cost price of imported steam coal.

The first is to start with the current price and **link its future movement** more or less directly to **that of fuel oil** or crude oil. In this case, the price of coal is aligned in the relatively long term to that of oil, less a certain margin. It should be noted that until very recently the price differential in favour of coal has been over $85/toe and that, power stations and cement factories apart, there has been no significant move in industry towards conversion to coal.

The second method – which has been selected here – is to take the cost price of the **marginal supply source** of coal imported into the Community as the trend indicator: in this case, coal from the east coast of the USA and from Australia (New South Wales).

In both cases, the average production cost ex-pit in the installations using the best technologies was estimated at $48/t. Inland transport and port costs increase this by $12.5/t in the USA and $9.5/t in Australia; the cost of shipping the coal to Europe is $8.5/t and $20/t respectively. Thus, the cost per marginal tonne of steam coal cif ARA lies between $69 and $77.5/tce.

The **structure of production costs** in the coal industry is dominated, as a rule, by investment and labour. This implies:

(a) significant borrowings on the capital markets, where interest rates may well remain relatively high in real terms; and

(b) faster-than-inflation increases in the average wage, mainly because of the need to employ increasingly skilled labour.

These two elements explain a drift in production costs above inflation, even in the USA, on an underlying trend estimated at about 2% per annum.

The average cost price per tonne of steam coal imported into the Community could therefore tend eventually towards the cost of such marginal production. The resulting central trend assumption is shown in the table below.

Central trend assumption for the average cif cost price of steam coal imported into the Community
$US at 1983 prices

	1975	1980	1981	1983	1990	1995	2000
$US/tce	68.5	65.5	71.0	57.5	85.0	95.0	105.0

16.   The first method was used, however, to determine the **cost price of imported steam coal** corresponding to the two alternative trend scenarios for the price of oil.

In the **"high" scenario,** the price of steam coal was aligned on the price of oil (less a certain margin) as long as the competitiveness threshold of the new gas-producing areas was not reached.

Once this threshold was reached, the price of coal was then aligned on the procurement price of gas, less an amount representing the additional investment required if coal is to be used instead of gas for raising steam.

In the **"low" scenario,** the price of coal was aligned to the price of oil for as long as possible, but without going below the average cost price per marginal tonne imported from the USA.

The results of these two sets of alternative assumptions are given in the following table:

Two alternative trend scenarios for the cost price of steam coal imported into the Community
$US at 1983 prices

$US/tce	1980	1981	1983	1990	1995	2000
High scenario	65.5	71.0	57.5	120.0	150.0	150.0
Low scenario				76.0	84.0	93.0

17.   To these different assumptions for the cost price of imported steam coal, there now only needs to be added:

- either: the price differential for the quality of the product, to obtain the equivalent import prices for coking coal and house coal;

- or: the average subsidy level permitted by each producing country in the Community, in order to obtain the costs for coal produced by each country.

By weighting these prices by the relative share of each national market covered by national output or by imports we obtain the average supply price for each type of coal.

National production costs for peat and lignite were not studied in detail. For the purposes of the study they were simply taken as equivalent to 75% of the price of imported coal on the basis of their energy content.

b) Consumer coal prices

18. As with the other energy products, two components were added to obtain the prices of coal to consumers: distribution costs and VAT.

The first component represents the inland transport and distribution costs in each country, which vary according to sector, structure of supply and the remoteness of the consumer areas from the areas of supply.

These costs were taken as constant over the period but varying from one Member State to the next.

The assumptions about the second component, VAT, are the same as for oil and gas: current rates continue to apply until the turn of the century.

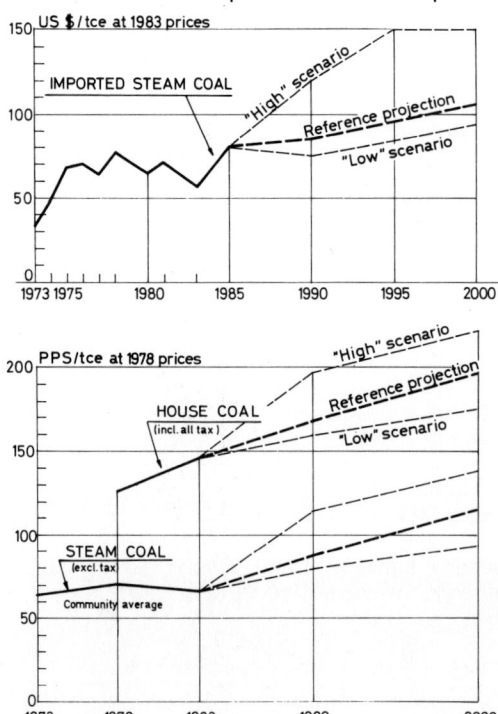

EUR-10:COAL

Trend assumptions for consumer prices

V.   Electricity

a)   Cost of electricity generation

19.   There are three components in the **production cost** of electricity:
      depreciation of fixed assets, maintenance and operating costs, and
      fuel costs.

      These three factors, although easy to measure at each plant, are
      difficult to determine in practice at macroeconomic level since they
      are strongly dependent on such things as changing production
      capacities, fuel used, the age of the plant, and daily and annual
      operating periods.

      A recent Commission in-house study on nuclear energy[7] nevertheless
      contains some information on the levels of these three factors in the
      case of new nuclear and coal-fired power stations in certain
      conditions of use at the beginning of the 1990s.

      This reveals that, given the assumptions,[8] **the average cost** of
      nuclear-produced current would be 31 ECU (1983) per MWh produced, the
      values for the individual countries ranging from 20% below to 40%
      above the average. The Community average production cost for
      coal-produced current would be 57 ECU/MWh, the values for the
      individual countries ranging from 40% below to 10% above the average.

      Although the deviations from the mean are considerable, reflecting
      each Member State's particular situation, the national ratios of
      coal-fired to nuclear generating costs lie within fluctuation margins
      of approximately 20% round the Community average.

      Recently published data on one Community country[9] shows that with an
      annual utilization of between 4 000 and 8 760 hours/year, the cost of
      generating electricity from fuel oil would be twice as high as from
      coal, desulphurizing costs included in both cases.

      The unit cost of hydro-electricity generation was taken to be
      13.5 ECU/MWh in all Member States. This is accounted for by average
      depreciation alone, since the other cost components are negligible.

20.   Weighting unit costs by type of power station, allowing for the
      predicted trend in the structure of electricity generation and the
      combined assumptions for the rate at which power stations are
      decommissioned and replaced and capacity expanded, we made an
      approximate calculation of the long-term trend of electricity
      generating costs.

----

[7] Nuclear Indicative Programme, Commission document COM(84)653 of
    22 November 1984.
[8] A discount rate of 5%; 6,600 hours utilization a year; new power
    stations scheduled to enter into service around 1990.
[9] Data published by the French Ministry for Industry and Research in
    April 1984.

But given the considerable - and not entirely foreseeable - lead times between the decision to invest in new plant and entry into service, the relatively long life of electrical equipment in general and the very wide range in Member States' capital unit costs and operating costs, it seemed risky to extrapolate the medium-term trends of these two electricity generation cost components. It was agreed therefore to keep them fixed, in real terms, throughout the projection period.

Long-term generating costs will thus vary, up to the year 2000, only on the basis of the trend assumptions selected for fuel prices and the proportion of total electricity generation accounted for by each generating sector.

The assumptions selected for the projected trend, by country, of conventional fuel prices are described above. For nuclear fuel, a drift of 1% above inflation for the whole period was assumed.

The trend of the average cost of electricity generation resulting from the assumptions selected is summarized in the table below. The Community trend is obtained from the weighted average of generating costs in the Member States.

Central projection of the trend in the average cost of
electricity generation in the Community

$US per MWh at 1983 prices

	1980	1983	1990	1995	2000
($US/MWh)	60.0	50.0	51.0	52.5	54.0

21. Using the same set of assumptions but the high and low scenarios for fossil fuel consumer prices, we can calculate the corresponding average cost of electricity generation. For the Community as a whole these alternative average costs are as follows.

Two alternative trend scenarios for the average cost of electricity
generation in the Community

$US per MWh at 1983 prices

$US/MWh[10]	1980	1983	1990	1995	2000
High scenario	60	50	61	65	61
Low scenario			48	48	48

---

[10] For calculating the production equivalence, 4.55 MWh approximately = 1 toe, i.e. 1 kWh = 2 200 kcal.

b)   Consumer prices

22.   Once the trend of average generating costs is determined,   only an estimate of the costs of transport and distribution needs be added in order to establish the prime cost of the current.

By comparing the **cost price of electricity** with the average prices charged to various consumers,   it was possible to calculate a theoretical,   approximate structure for electricity margins.   The calculation for recent years shows a relatively similar distribution structure for all Member States.

Ranges for the average gross margins on cost prices of electricity
distributed to the "typical" consumers selected (EUR-10)

Small industrial consumer	(1.25 GWh/yr)	1.1 to 1.2
Large industrial consumer	(10 Gwh/yr)	0.9 to 0.95
Small domestic consumer	(3.5 MWh/yr)	1.3 to 1.5
Large domestic consumer	(20 MWh/yr)	1.0 to 1.1

The **average gross margins** were then used to calculate an average price (exclusive of tax) for electricity sold to consumers, which was consistent with the various assumptions about fuel prices and, upstream, the cost of the crude oil supply.

To find the **real average price of electricity paid by the consumer,** we simply have to add in excise duties and any other taxes and VAT.

As with other energy sources, it was assumed that taxes will continue to be levied at their current level until the end of the century. Neither excises nor any other taxes, except VAT, are generally levied on electricity except in three countries: Germany, Italy and Denmark.

EUR-10: ELECTRICITY

Trend assumptions for average consumer prices

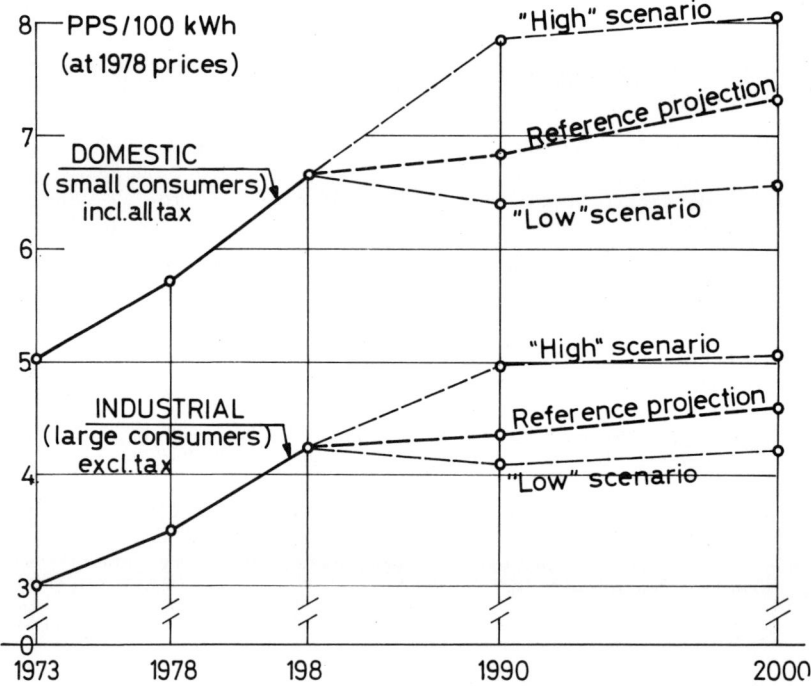

Annex 1: Summary graphs

EUR-10: Purchase costs of imported fossil fuels

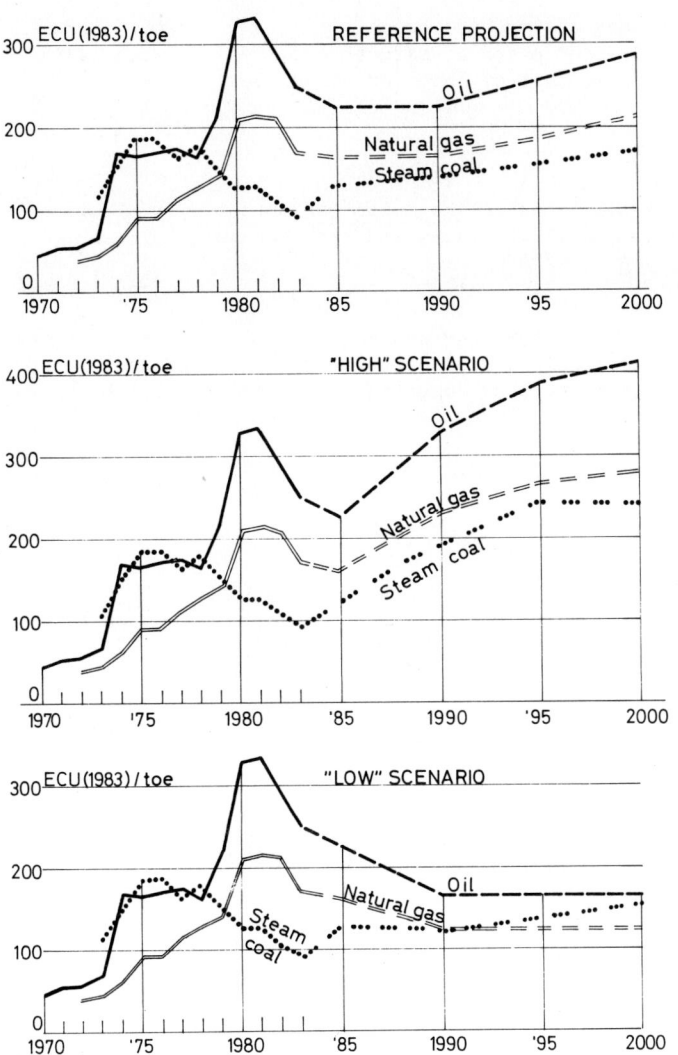

Annex 2: Average energy purchase costs for the Community

(a) Assumptions in the reference projection
    (ECU per toe, at 1983 prices and exchange rates)

EUR-10	1983	1990	2000
Average cif cost of imported crude oil	248.5	225	290
Average procurement price of natural gas	170.5	165	215
Average cif prime cost of imported steam coal	92.5	135	170

b) Extreme assumptions in the price variants
   (ECU per toe, at 1983 prices and exchange rates)

EUR-10	1990	2000
Average cif cost of imported crude	330-165	415-165
Average procurement price of natural gas	235-125	280-125
Average cif prime cost of imported steam coal	195-120	240-150

C.   THE COMMUNITY'S ENERGY SUPPLY IN THE LONG TERM

I. Energy supply in 1990

1.   The outlook for Community supply depends on technical, economic and
     political conditions in the Community and on the world energy market.
     Only the political conditions in 1990 are completely unpredictable.
     The technical and economic conditions can be estimated reasonably
     without risk of a major error, since they depend at this relatively
     near point in the future on supply infrastructures that either exist
     already or are in the process of being built and on supply contracts
     that have generally either been concluded or are being negotiated.

2.   A recent Communication[11] from the Commission to the Council examines
     the likely major supply flows for the Community in 1990.   The table
     below summarizes the projections to 1990 contained in this document.

Community Energy Supply - projections to 1990

Million toe	Community production		Imports from non-Community countries	
	1983	1990	1983	1990
Coal	143.1	141.5	38.5	55.4- 70.4
Lignite and peat	31.0	36.5	0.4	..
Oil	132.5	103.2-108.2	288.8	333.1-342.1
Natural gas	119.9	113.0-114.5	48.212	69.5- 73.0
Nuclear	76.1	135.0-148.5	1.9	(-1.6)-(-3.3)[12]
Hydro & geothermal	12.0	13.0	..	..
New energy sources	1.7	8.9-11.9	..	..
Total	516.3	551.1-574.1	377.9	456.4-482.2

3.   As regards **primary energy production** for the whole period 1983-1990,
     these projections reveal:

     - relative stability in coal-mining and hydro-electric power; the
       maintenance of coal production at present levels may be regarded as
       a maximum limit, given the high grants and subsidies needed to make
       continental Community production competitive with imported coal.

     - the expansion of lignite and peat extraction owing to the
       development of production capacity in Greece;

---

[11] Review of Member States' Energy Policies - COM(84)88 final,
     29 February 1984.
[12] Net exports of electricity.

- the appearance of new and renewable energy sources, mainly the result of a data-recording change in one Member State (the inclusion of firewood);

- the large growth in nuclear electricity generation reflecting the doubling of operational capacity in the Community (100 GWe by 1990), half of it in France;

- a slight contraction in natural gas production due mainly to the progressive exhaustion of the French and Dutch fields; but Dutch production in 1990 could be kept at 1982/83 levels, thus stabilizing Community production overall at its present ceiling;

- a significant reduction in oil extraction, mainly in the UK sector of the North Sea. These projections, however, may prove too pessimistic and changes in the UK tax system since they were made were intended to encourage the exploitation of fields which are today regarded as marginal.

4. **Energy imports** from non-Community countries depend on aggregate levels of demand and internal production. Nevertheless, a few important trends emerging from the projections to 1990 should be stressed:

- coal imports are expected to increase by between 40 and 80% in seven years; this implies substantial power station conversion from oil in most Member States and a substantial return to coal in certain industries;

- a barely less significant increase in natural gas imports (45-50%) as the contracts concluded take effect, which implies increased conversion from oil, especially in the residential and tertiary sector, and in industry in some Member States;

- oil imports increase slightly to return to 1982 levels; as oil remains the most abundant and most flexible marginal source of energy supply, oil imports could increase more substantially if coal were not to achieve the expected breakthrough on industrial markets around 1990;

- the Community becomes a net exporter of (mainly nuclear) electricity.

II. Energy supply in the year 2000

5. Despite the rigidities of the energy market, there is still a good deal of uncertainty as to the share which each source of energy will hold on the Community market in 1990. Such uncertainty is of course even greater for the year 2000.

The experience of the years 1974-1984 has shown that major structural changes could occur in less than ten years.

It therefore seems preferable simply to set, in broad outline, the limits between which the Community's energy supply could fluctuate in 2000.

6. This approach, identifying conditions on the international market and the particular characteristics of the Community market, was adopted for the scenario study to the year 2000 carried out by the Commission's departments in 1983.[13]

    This initial approach was supplemented by:

    (a) a specific study of the world energy market in the year 2000 (see Chapter VI), and

    (b) several recent sectoral papers sent by the Commission's departments to the Council concerning:

    - the natural gas market;[14]

    - the potential share of solid fuels on the energy market and the competitiveness of the solid fuel industry in the Community;[15]

    - the illustrative nuclear programme for the Community.[16]

(a) Natural gas supply

7. Within the Community, production of natural gas in 2000 should be between 90 and 110 million toe, failing any major new discovery.

    As production from the Groningen field has started to decline, the Netherlands will produce no more than 40 million toe of natural gas a year in 2000, as against 55 million toe today. In the United Kingdom, production will be between the current level of 30 million toe and a ceiling of 40 million toe.

8. The expansion of international trade in natural gas during the last ten years has led to extremely expensive investments in transport infrastructure with a long useful life. Normally, contracts involve long-term offtake commitments. Some contracts concluded will therefore still be running in the year 2000. Thus, on the basis of existing contracts, the Community is already bound to import 6 million toe from Norway, 40 million toe from the USSR and 20 million toe from Algeria.

---

[13] See the Commission staff paper "Scenarios to the year 2000 - a prospective study of the energy market in the European Community", 1983.
[14] COM(84)120 final, 13 April 1984; COM(84)583, 26 October 1984.
[15] SEC(83)1925 and 1926, 22 and 23 November 1983.
[16] COM(84)653 final, 22 November 1984.

9. Various extra possible sources of natural gas supply could become available during the 1990s. These are uncertain, because they will depend on the assessment which Community gas distributors make of the possible growth in demand and hence of the long-term commitment they are willing to make as regards offtake levels and purchase prices. These additional deliveries could come either from now normal suppliers (Norway, USSR and Algeria) or from new producers such as Nigeria, Cameroon, the Ivory Coast, Canada, Qatar or Abu-Dhabi. An additional 60 million toe approximately of gas could be available for Western Europe in the year 2000.

Assuming that the Norwegian Troll field is brought to production in the middle of the 1990s, an additional volume of gas of about 35 million toe could be added to this figure.

The Community's possible sources of natural gas supply

EUR-10 (million toe)	1983	2000
Internal production	120	90 - 110
Imports covered by contracts	50	77 [17]
of which Norway	20	17
Algeria	12	20
USSR	18	40
Further possible imports		
Norway		15 - 50
USSR		20
Algeria		15
Nigeria		5 - 10
Others		p.m.
Total possible supply	170	222 - 282

(b) The supply of solid fuels

10. Community coal prices are generally aligned on the prices of imported coal, around $60 per tonne in 1983. As coal production costs in the Community are higher the losses incurred amounted to between $17 and $26 per tonne produced. The level of Community coal production in the year 2000 will thus depend both on the ability of mining undertakings to improve their financial situation and become more competitive against imported coal, and on the prolongation of a stable aids policy which will continue to allow Community deposits to be worked under agreed rules.

It can be assumed that 15% of Community production is completely uneconomic, and the eventual closure of that uneconomic capacity is to be expected. As 20% of Community production can easily meet competition from imported coal, there remains 65% of the coal industry which can be described as marginally non-competitive.[18] It

---
[17] Including Sleipner.
[18] COM(82)31 final, 10 June 1982; SEC(83)1925, 2 December 1983.

is difficult to predict what proportion of this will improve efficiency enough to become competitive and what proportion will continue to be subsidized. In the reference projection it is assumed that the larger part of non-competitive production will be preserved and that the effect of closures will be offset by increases in output resulting from the development of new pits and increased productivity[19] in older workings. On these assumptions, Community coal production should decline from its 1983 level of 140 million toe to approximately 131 million toe in 2000.

11. It is predicted that Community lignite and peat production will expand by 1990; as no further changes in production capacity seem to be contemplated thereafter, output in 2000 is likely to be at its 1990 level, i.e. approximately 35 million toe.

12. Community coal imports should grow appreciably in order to meet additional demand, in the main from power stations. As explained in the chapter on world supply, coal production in the industrialized countries should go up significantly by 2000, especially in the USA, Australia, Canada and South Africa. These countries therefore, and possibly Poland too, could increase their exports to Europe.

All things considered, coal imports into the Community could go up from 40 million toe in 1983 to 100 million toe or even more in 2000. The level will depend ultimately on the extent to which coal penetrates the industrial heat market.

Nearly 50% of coal imports entering the Community in 2000 could come from the USA, one quarter from South Africa, and the remainder from Canada, Poland, Australia and perhaps China.

Potential Community supplies of solid fuel

EUR-10 (million toe)	1983	2000
Internal production:		
- coal	143	120 - 140
- lignite and peat	31	35
Potential coal imports from non-member countries:	41	75 - 145
USA	17	35 - 55
South Africa	10	20 - 30
Canada	1	5 - 10
Poland	7	5 - 10
Other East European countries	1	0 - 10
Australia	6	10 - 20
China and the Far East	p.m.	0 - 10
Potential total supply	215	230 - 320

----

[19] Excluding recovered products.

(c) Nuclear energy

13. During the next fifteen years nuclear energy will continue to
     contribute to the growth in electricity generating capacity. The
     foreseeable trend in electricity demand is therefore a decisive
     factor in nuclear industry investment planning.

     The uncertainty is less for the period up to 1995, as the decisions
     on investments to be made by that date have had to be practically
     finalized this year. It is estimated that installed nuclear capacity
     in the Community in 1995 will be 120 GWe, which represents an
     increase of about 25 GWe on 1990. It therefore seems reasonable for
     the reference projection to include a potential increase of
     comparable scale between 1995 and 2000.

Potential Community supply from nuclear sources

EUR-10	1983	2000
Installed nuclear capacity	52 GWe	145/150 GWe
Nuclear generated electricity	295 TWh	900/1000 TWh

(d) Oil supply

14. Current and planned levels of activity in the Community as regards
     exploration and bringing into service suggest that Community
     production has reached a plateau today which may continue for a few
     years, before falling slowly to a level of approximately
     110 million toe in 2000.

15. Community oil requirements could level off at about
     400/500 million toe during the next fifteen years, which would mean
     imports of 300/350 million toe in 2000.

     The analysis of world energy prospects given at the end of this
     paper   and more precisely the part devoted to the supply of oil,
     indicates that, failing unpredictable major upsets, availabilities in
     the traditional oil-exporting countries will make it possible to meet
     Community demand as well as that of the rest of the world.

     It is assumed, therefore, that in the year 2000, the non-OPEC
     producer countries (e.g. Norway and Mexico) will supply to the
     Community between 65 and 130 million toe of crude oil and petroleum

products. In this event, exports from OPEC countries to the Community would be a little above their 1983 level, about 220 million toe per annum.

The Community's potential oil supply

EUR-10 (mtoe)	1983	2000
Internal production of crude oil	132	110
Imports of crude and petroleum products from:		
— OPEC	220	240 — 290
— USSR and Eastern Europe	60	40 — 60
— Other	50	25 — 70
Exports to non-member countries	42	..
Total potential supply	434[20]	405 — 520

(e) New and renewable energy sources

16. As most sites in Europe suitable for the production of hydroelectric power have already been developed, the level of hydro generation should remain much the same until the end of the century.

Other new and renewable energy sources should increase rapidly by 2000, but their share in meeting the Community's energy needs will remain marginal, at least as recorded in official statistics. Energy production from all new and renewable sources together would be equivalent to 20-30 million toe; one half of this would come from hydro.

However this traditional approach to the question underestimates the real contribution of new and renewable energy sources towards meeting the Community's energy requirements. These forms of energy are highly decentralized in the energy system and, being used directly by the producer, are for the most part not marketed and therefore do not appear in the statistics. Examples are passive solar energy, firewood. When these forms of energy do appear in the data, they are generally recorded in terms of calorific content rather than in energy replaced and in many cases this tends to understate their contribution to the energy balance.

---

[20] Includes 17 mtoe supplied from stocks.

EUR-10: Potential Energy Supply

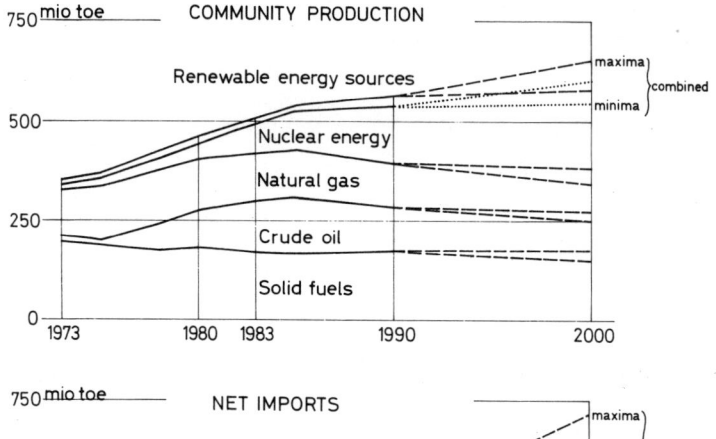

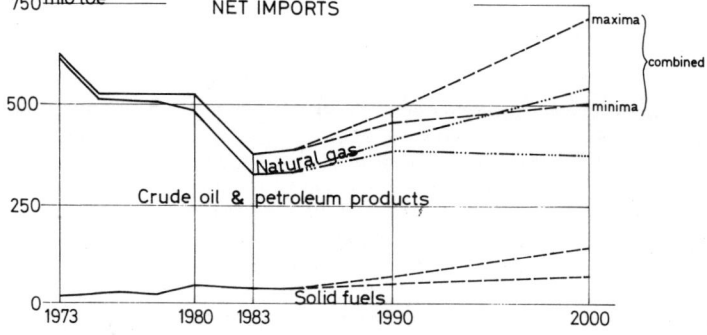

CHAPTER II

REFERENCE PROJECTION: Final Energy Demand

Taking the main assumptions about future economic trends (growth and structure of GDP), the prices of the various forms of energy and potential sources of supply, this chapter analyses the consequences for energy demand in each sector of final consumption: industry, transport, and residential and tertiary in the Community as a whole and in each Member State.

I. The overall picture

1. Between 1973 and 1980 final energy demand in the European Community remained relatively stable and then decreased by 3% per annum from 1980 to 1983. Over the same period, GDP increased by an average of 2.3% per annum until 1980, and then by 0.3% per annum during the next three years.

   The final energy intensity of the Community economy[1] thus showed a marked improvement, for in ten years it was brought down from 0.64 toe/1000 ECU to 0.49 toe/1000 ECU. This trend should continue, but at a slower rate, the final energy content of the Community economy decreasing to 0.46 toe/1000 ECU in 1990 and 0.38 toe/1000 ECU in 2000. All the consumer sectors should contribute to this trend, as indicated in the table below and the sectoral analysis which follows it.

2. By source of energy, final consumption of petroleum products should be fairly stable between 1983 and 2000, there should be a moderate increase for solid fuels, a bigger increase for natural gas and steady growth in the final consumption of electricity and of distributed heat.

   Thus, the proportion of final energy consumption covered by solid fuels will decrease from 12% in 1973 to 8% in 2000; over the same period, the share of oil will decrease from 62% to 47% and natural gas and electricity will increase their shares from 15 to 24% and from 11 to 18% respectively.

----

[1] The final energy content of the economy is calculated by dividing final energy consumption (including non-energy uses and all expressed in toe) by GDP, measured in ECU, at 1975 prices and exchange rates.

Final energy consumption in the European Community

million toe

EUR 10	1973	1980	1983	1990	2000	1990/82	2000/90
Industry	247.9	227.0	186.4	216	234	+2.1%	+0.8%
Transport	128.2	153.6	155.6	169	182	+1.2%	+0.7%
Domestic & Tertiary	264.9	265.8	258.6	267	274	+0.8%	+0.3%
Non-energy uses	70.3	60.0	57.4	62	72	+1.1%	+1.5%
Total	711.3	706.4	652.0	714	763	+1.3%	+0.7%
of which							
− solid fuels	85.4	63.1	55.3	58	61	+0.7%	+0.5%
− oil products	439.3	395.1	351.2	363	359	+0.5%	−0.1%
− gas	107.4	151.5	147.2	169	185	+2.0%	+0.9%
− electricity	76.0	92.8	94.6	113	138	+2.6%	+2.0%
− Other[2]	3.2	3.9	3.7	12	18	+18.3%	+4.1%

II. Industry

3. Most industrial energy consumption is concentrated in seven sectors.
The four most energy-intensive are iron and steel, non-ferrous
metals, non-metallic minerals and paper and pulp. They accounted for
15% of industrial value added in the Community in 1980 and absorbed
50% of the energy consumed by industry. Three other sectors –
chemicals, textiles and food processing – are also heavy consumers of
energy, accounting for 30% of energy consumed by industry and 30% of
industrial added value.

The assumptions in the reference projection allow for substantial
restructuring of these two groups of energy-intensive industries,
which will shift production to goods with higher added value.
Changes in the overall structure of industry will reduce their
contribution to industrial value added to some 40% by the year 2000,
as the emphasis moves towards light industry and capital goods.
These two changes together will substantially reduce the average
energy content of Community industrial production.

4. As regards energy uses, these shifts in industrial output will
encourage the use of low-temperature processes and of electricity for
specific purposes at the expense of high-temperature processes.

---

[2] Heat and renewable forms of energy.

Final energy requirements of
industries by type of use

EUR 10	1983	%	1990	%	2000	%
Steam	65	(34)	79	(36)	88	(37)
Furnaces	87	(45)	91	(41)	92	(38)
Motor fuels	4	(2)	4	(2)	4	(2)
Specific electricity	36	(19)	47	(21)	56	(23)
Total	192	(100)	221	(100)	241	(100)

The restructuring of the energy-intensive industries will be
accompanied by major changes in processes which will also lead to
energy saving, for example in the iron and steel industry (more use
of scrap and prereduced pellets), cement factories (steady decline of
the wet process), paper pulp (greater use of wood and biomass), and
petrochemicals (use of advanced crackers for the production of
ethylene).

5. As industry is being restructured, a greater emphasis on investment
    should speed up the modernization of fixed capital and improve its
    performance, including its energy efficiency.

    A clear trend towards mechanization, automation and better plant
    controls will also facilitate substitution between different forms of
    energy, and will generally lead to a sharper rise in electricity
    consumption.

6. These changes, generally favouring energy saving, will result overall
    in a major decrease in the unit consumption of energy in the
    Community's industries, which will drop from 0.44 toe/1000 ECU^3 of
    industrial value added in 1983 to 0.41 in 1990 and 0.36 in 2000.

7. Energy consumption by the Community's industries as a whole will
    increase by an average 1.3% per annum, from 186 million toe in 1983
    to 234 million toe in 2000. Over the same period, industrial
    production will increase at a rate of some 2.3% per annum.

    As economic recovery progresses, the consumption of petroleum
    products will increase in volume terms up to 1990 and then fall back
    to approximately 1983 levels by 2000, which will mean a significant
    decline in their share in final consumption by industry, from 26% in
    1983 to 22% in 2000. The shares of natural gas and solid fuels will
    remain stable up to 2000 despite some revival in the use of steam
    coal in industry as low-temperature final uses increase. Recovered

---

[3] The ratio between total energy consumption by industry and the value
added by the whole of industry at 1975 prices and rates of exchange.

steam will increase its market share, and so will electricity, its share of the industrial market growing at an average annual rate of 2.1%.

Industrial energy consumption by fuel

EUR 10	1973	1980	1983	million toe 1990	2000
Solid fuels	45.6	38.4	35.7	40	46
Oil products	104.7	77.6	49.3	57	52
Gas	57.9	66.8	59.6	67	73
Distributed heat	1.1	1.4	1.2	4	5
Electricity	38.6	42.8	40.6	48	58
Total	247.2	227.0	186.4	216	234

III. Transport

8. In 1983 the transport sector as a whole consumed 156 million toe of energy, of which 85% was accounted for by road transport, 10% by air, 3% by rail and 2% by inland waterway transport.   Overall, approximately 60% of the traffic was the carriage of passengers and 40% the carriage of goods.

9. As in the past, the growth of the **goods transport sector** will be closely linked with economic growth in general and growth in the productive sector in particular, which will average 2.3% per annum during the period 1983-2000. However, in the reference projection, the shift in industrial production to high added value products together with technical improvements to the vehicle fleet will increasingly decouple the two trends. Energy consumption in the goods transport sector may be expected therefore to grow at around 1.5% p.a. between now and 2000.

10. Developments in the **passenger** sector are more difficult to discern insofar as changes here will depend on changes in the stock of vehicles, their average specific consumption, annual mileage, urban/inter-city proportions and the split between private and public transport. In the reference projection, between 1983 and 2000 the stock of private cars in the Community will probably increase by some 50%. On the other hand, the average distance covered annually per car will decrease by about 20% in most Community countries, while average consumption per 100 km will decline by 15 to 20% in both urban and inter-city traffic. Public transport will slightly increase its share in both urban and inter-city traffic.

The combination of all these factors will probably lead to a relatively modest growth in the energy requirements of the passenger transport sector, set to increase by 0.4% per annum up to 2000. By 2000, passenger traffic will accordingly account for only 55% of total energy consumption and the carriage of goods will account for the remaining 45%.

11. Energy consumption in the transport sector per unit of GDP[4] is therefore projected to decline from 0.12 toe/1000 ECU in 1983 to 0.11 in 1990 and 0.09 in 2000. The transport sector's total energy consumption will then increase from 156 million toe in 1983 to 182 million toe in 2000, giving an average rate of increase of 0.9% per annum over the whole period.

Electricity consumption will increase at the same rate as the growth of rail transport, i.e. by 2.5% per annum. Other energy requirements in transport will be met by fuels for road vehicles. Here the following should be noted:

- the rapid growth of LPG (+5% per annum), whose share of the total market for fuels will nevertheless remain below 4% in 2000;

- a slight increase in the share of gas oil, from 31 to 38% between 1983 and 2000; this will result from an increasing switch to diesel fuel in private cars and the expansion in road haualge.

Energy consumption in the transport sector
by type of fuel

million toe

EUR 10	1973	1980	1983	1990	2000
Oil products	125.1	150.7	152.8	166	178
of which – petrol	75.4	84.5	87.0	85	78
– gas oil	34.2	44.9	48.0	58	70
– LPG	0.8	1.5	2.4	4	6
Natural gas	0.3	0.3	0.3	0	0
Electricity	2.0	2.4	2.4	3	4
TOTAL	128.2	153.6	155.6	169	182

IV. Residential and tertiary sector

12. Between 1983 and 1990, private consumption in the Community will probably increase by an average of 1.3% per annum, compared with an increase of 0.8% in the energy consumption of the residential and tertiary sector. Between 1990 and 2000 the two rates will be increasingly decoupled as private consumption increases at an average

---

[4] Energy consumption in the transport sector divided by GDP in ECU, at 1975 prices and exchange rates.

rate of 2.7% per year and the private sector's energy requirements rise by only 0.3%. Over the period to 2000 total energy consumption by the residential and tertiary sector should therefore rise by an annual average rate of 0.6%. Energy consumption of the sector per unit of GDP declines from 0.193 toe/1 000 ECU in 1983 to 0.173 in 1990 and 0.136 in 2000.

This total masks differences in user trends. For the Community as a whole, energy consumption for space heating will probably decrease at a steady rate throughout the period, whereas there will be a steady increase in consumption for the production of domestic hot water, for cooking and for specific uses of electricity.

13. The main energy application in households is space heating. This currently accounts for approximately 70% of the final energy consumption of the sector, i.e. some 170 million toe in 1983.

Energy consumption for space heating in 2000 will depend on trends in various determining factors, including:

- the number of housing units, and the proportions of renovated dwellings and new dwellings;

- the structure of the housing stock and of the heating equipment used;

- the degree of insulation in buildings and the unit output of equipment.

Generally speaking, these factors should run on convergent lines up to 2000 and bring about a general decrease in the consumption of energy for heating purposes.

14. In the Community as a whole, population growth should remain moderate, increasing from 272 million in 1983 to 280 million in 2000. In the reference projection the average number of persons per household will remain relatively stable throughout the period, and therefore the number of households and the number of housing units should increase at an average rate of 0.2% per annum up to 2000.

It may be assumed that in 2000 approximately half of the 108 million housing units will be either new or renovated, or housing in which central heating systems have been substituted for individual systems. The use of central heating equipment should spread so that, by 2000, it will account for an average 85% of all heating equipment. The average will be slightly higher for the countries of northern Europe.

At the same time, there will be a 20-30% improvement in the insulation of the buildings between now and the year 2000, as a result of stricter standards for new buildings and improved insulation of existing dwellings. Finally, technical improvements in individual heating equipment and especially central heating plant will make for substantial improvements in equipment efficiency.

The combination of these various factors will lead to an average 15% decrease in the amount of energy consumed per housing unit in the Community between 1980 and 1990. This unit consumption should further decline by a similar amount in volume terms during the 1990s. The general trend will be common to all the countries, although in some it will be much more marked.

15. Energy requirements for the production of domestic hot water in the residential sector will increase by an average 1.2% per annum during the period; but in the tertiary sector, energy requirements for the production of hot water will remain almost unchanged. Community energy consumption for cooking will probably increase by 0.8% per annum between now and the end of the century.

16. Alongside these moderate increases in the calorific requirements of the residential and tertiary sectors in the Community between now and 2000, there will be a relatively steady increase of 2.9% per year in the consumption of electricity for specific uses, such as motive power, household appliances and lighting.

Final energy requirements of the residential
and tertiary sector,[5] by use

	million toe					
EUR 10	1983		1990		2000	
		(%)		(%)		(%)
Space heating	169	(70)	168	(66)	158	(61)
Domestic hot water	37	(15)	41	(16)	45	(18)
Cooking	7	(3)	8	(3)	9	(3)
Specific electricity	28	(12)	35	(15)	46	(18)
TOTAL RESIDENTIAL AND TERTIARY	241	(100)	252	(100)	258	(100)

17. By the end of the century in the Community as a whole, the sector will show substantial penetration by natural gas (+30%) and electricity (+50%) for thermal uses and distributed heat. These forms of energy will therefore partly supplant coal and oil products, consumption of which will decrease by 60% and 20% respectively during the period.

---

[5] The data do not include the agricultural sector, whose total consumption will probably increase from 11 to 16 million toe during the period under review.

Energy consumption in the residential and tertiary sector
(including agriculture), by type of fuel

million toe

EUR 10	1973	1980	1983	1990	2000
Solid fuels	35.4	21.8	18.2	16	13
Oil products	148.8	117.4	101.3	90	71
Gas	43.2	76.5	79.0	92	100
Distributed heat	2.1	2.5	2.5	8	13
Electricity	35.4	47.6	51.6	61	77
TOTAL	264.9	265.8	252.6	267	274

V. Non-energy uses

18.   The main non-energy uses cover oils and lubricants, bitumens and
waxes and oil products (in particular naphtha) and natural gas used
as petrochemical feedstocks. The use of products for purposes other
than the production of energy is therefore mainly limited to motor
vehicle or motive power service activities, public works and the
chemical industry.

19.   Between 1983 and 2000 GDP will probably increase by 2.6% per annum on
average, while energy consumption for non-energy uses will probably
increase by 1.3% per annum during the same period, without any
significant dent in the dominant share held by petroleum products.

The consumption of coal for non-energy purposes will increase
slightly, although the coal-chemical industry will remain fairly
marginal.

Consumption of energy products for non-energy purposes

million toe

EUR 10	1973	1980	1983	1990	2000
Solid fuels	3.4	2.7	1.3	2	2
Oil products	60.7	49.4	47.8	51	58
Natural gas	6.2	7.9	8.3	10	12
Total	70.3	60.0	57.4	63	72

★

★   ★

CHAPTER III

REFERENCE PROJECTION: Energy supply

With the energy requirements of the final consumption sectors determined, this chapter identifies the optimal primary energy supply structure to meet that demand at least cost, taking into account the constraints on available supplies in 1990 and 2000.

I. Trends in gross inland energy consumption

1. The table below shows the pattern of Community primary energy demand related to the reference projection for final energy demand, and the breakdown of the requisite supply between Community production and imports.

Summary energy balance – European Community (EUR 10)

million toe

	1973	1979	1983	1990	1995	2000
I. Gross energy consumption	968.0	1011.7	907.2	1034	1086	1136
– bunkers	37.3	26.9	22.0	28	28	29
– internal consumption	930.7	984.8	885.2	1006	1058	1107
II. Internal consumption	930.7	984.8	885.2	1006	1058	1107
– solid fuels	222.0	223.4	212.0	242	255	264
– petroleum products	563.9	536.6	416.3	413	413	410
– gas	115.8	172.4	165.3	190	190	196
– primary electricity etc.	29.0	52.4	91.6	161	197	236
III.Internal production	351.3	457.9	516.3	563	591	625
– coal	171.1	149.2	143.0	139	136	135
– lignite and peat	26.5	31.0	31.0	36	36	38
– oil	13.2	89.3	132.5	111	111	108
– natural gas	112.2	137.5	119.9	115	111	108
– nuclear energy	17.7	37.2	76.1	145	178	215
– hydro & geothermal	9.4	12.2	12.0	13	14	14
– other and renewables [1]	1.2	1.5	1.7	4	5	7
IV. Net imports	619.9	558.7	377.9	471	496	511
– solid fuel	19.0	33.8	39.1	67	82	92
– oil	596.2	487.2	288.8	330	330	330
– natural gas	4.0	36.2	48.2	74	83	88
– electricity	0.7	1.5	1.9	-1	–	–
V. Changes in stock levels	+3.2	-4.9	-13.1	..	..	

2. From 1963 to 1973 the Community's gross energy consumption increased by an average rate of 4.7% per year. The rate of increase then followed a jagged trend, returning in 1980 to a level close to that of 1973.

----

[1] For the sake of consistency with the historical data, the projections for renewable energy sources relate only to solar, wind and geothermal energy and the recovery of energy from domestic waste. The addition of wood and biomass could increase demand by 8 to 10 m toe in 1990 and 25 to 30 m toe in 2000.

Including all forms of consumption, gross energy consumption[2] in the Community is projected to increase by an average rate of only 1.7% per year until 1990 and then by 1% per year over the following decade, making an annual rate of increase between now and 2000 of 1.3%.

The moderate long-term increase in the Community's energy demand reflects not only the assumed moderate rate of economic growth but also structural changes in the economy and the results of effective energy saving policies.

From 1973 to 1983 energy intensity improved by about 20% in the Community. A further improvement on the same scale is expected between 1983 and 2000. According to the assumptions of the study, the primary energy intensity of the Community economy[3] should decline slightly from 0.67 toe/1000 ECU in 1983 to 0.64 in 1990 and then decrease more sharply to 0.54 in 2000.

The primary energy intensity of the Community economy
(toe/1000 ECU)

1973	1979	1983	1990	1995	2000
0.833	0.760	0.667	0.642	0.590	0.539

This trend is the result of the structural changes expected, as well as of investments in new energy saving processes and plant modernization which will improve energy performances.

3. Taking into account supply and demand in the electricity sector (see below), the relative shares of the different energy sources in the Community's gross consumption are expected to develop as follows:

- the share of solid fuels will remain stable;

- the share of oil will fall sharply to below 40% in 2000;

- the share of nuclear energy will increase substantially to meet 18% of energy requirements by the end of the century;

- the share of natural gas will increase slightly up to 1990 and then level off.

---

[2] Including bunkers.

[3] The primary energy intensity of the economy is calculated by dividing inland energy consumption (excluding bunkers) by GDP in real terms at 1975 prices and exchange rates.

The Community's gross energy consumption (including bunkers)

million toe

EUR 10	1973 (%)	1980 (%)	1983 (%)	1990 (%)	2000 (%)
Solid fuels	222.0(23)	222.7(23)	212.2(23)	242 (23)	264 (23)
Oil	601.3(62)	520.0(54)	438.3(48)	441 (43)	439 (39)
Natural gas	115.8(12)	169.3(17)	165.2(18)	190 (18)	196 (17)
Nuclear Energy	17.7 (2)	42.7 (4)	76.1 (9)	145 (14)	215 (19)
Hydro, geoth.& other	11.0 (1)	15.4 (2)	15.5 (2)	12 (2)	21 (2)
Total	967.8(100)	970.1(100)	907.3(100)	1034(100)	1136(100)

II. The reference scenario for energy supply

(a) Trends to date

4. From 1963 to 1973 the Community's inland energy production held
    steady at about 350 million toe, the increase in the production of
    natural gas offsetting the decline in coal output. Increased demand
    was therefore entirely covered by higher imports, which rose from 260
    million toe in 1963 to 620 million toe in 1973. Oil accounted for 95%
    of these imports.

    After 1973 the trends were completely reversed: by 1983 Community
    production amounted to 516 million toe and net imports to only
    378 million toe, of which oil accounted for no more than three
    quarters.

    The Community's dependence on external supplies therefore rose from
    43% to 64% over the period 1963-73 and then fell to 42% ten years
    later.

(b) Energy production in the Community

5. Nuclear energy is the only form of energy whose output will increase
    significantly in the Community by 2000. The reference projection
    assumes that the output of solid fuels will remain at a level close
    to that of 1983, and that of oil and natural gas will steadily
    decline, this projection being based on the fields now being worked
    or likely to be brought on stream.

Energy production in the Community

million toe

EUR 10	1973 (%)	1980 (%)	1983 (%)	1990 (%)	2000 (%)
Coal	171.2 (48)	153.3 (33)	143.0 (28)	139.0 (25)	135 (22)
Lignite and peat	26.5 (8)	31.8 (7)	31.0 (6)	36.0 (6)	38 (6)
Oil	13.1 (4)	91.1 (20)	132.5 (26)	111.0 (20)	108 (17)
Natural gas	112.2 (32)	129.2 (28)	119.8 (23)	115.0 (21)	108 (17)
Nuclear	17.7 (5)	42.7 (9)	76.1 (15)	145.0 (26)	215 (35)
Hydro & geothermal	9.1 (3)	12.3 (3)	12.0 (2)	13.0 (2)	14 (2)
Others & renewables	1.2 (0)	1.7 (0)	1.7 (0)	3.0 (0)	7 (1)
Total	353.2(100)	462.1(100)	516.1(100)	563.0(100)	625(100)

The figures given above for the Community's energy production in 1990 are very close to the projections notified by the Member States at the time of the last review of national energy programmes.[4]

The figures adopted for the reference projection for the year 2000 were obtained by combining a series of assumptions which are now regarded as plausible. But there are three major factors of uncertainty:

- the tonnage of new competitive coal-mining capacities which will come into production and without which output levels will decline sharply in the 1990s;

- the precise course which the nuclear energy industry will take in certain countries;

- the precise outlook in the late 1990s for oil and gas production in the Community, given the possibility - perhaps remote - of unexpected new discoveries of a major oil or gas field.

(c) Energy imports from non-Community countries

6. In 2000 the Community's energy imports should return to a level close to that of 1979, i.e. about 530 million toe. After falling to their lowest in 1983 (378 million toe), imports will rise steadily up to the year 2000 for each fuel: coal, oil and natural gas.

Net energy imports into the Community

million toe

EUR 10	1973 (%)	1980 (%)	1983 (%)	1990 (%)	2000 (%)
Solid fuels	19.0 (3)	47.3 (9)	39.1 (11)	67 (14)	92 (18)
Oil	596.2 (96)	437.9 (83)	288.8 (76)	330 (70)	331 (65)
Natural gas	4.0 (1)	40.6 (8)	48.2 (13)	74 (16)	88 (17)
Electricity	0.7 (0)	1.4 (0)	1.8 (0)	-	-
Total	619.9(100)	527.2(100)	377.9(100)	471(100)	511(100)

By the turn of the century imports will have been diversified in two ways: first, the shares of oil and other fuel imports will be better balanced, moving from 83% and 17% respectively in 1980 to 65% and 35% in 2000; and secondly, there will be a wider geographical range of supply sources.

7. The Community's dependence on oil imports for its energy supply will improve markedly over the next 15 years: from 62% in 1973 it has already been brought down to 32% in 1983 and should be further reduced to under 30% in 2000 if the basic assumptions in the reference projection are borne out.

Dependence on gas imports, however, will continue to rise until 2000. From nil in 1973 and 5% in 1983 it should increase to 8% in 2000 as consumption rises and Community production declines.

----

[4]Doc. COM(84)88, 29 February 1984.

For the same reasons dependence on coal imports will also increase, from 2% in 1973 and 4% in 1983 to 8% in 2000.

III. Main sectoral trends

(a) Oil

8. Between 1983 and 2000 oil consumption (including bunkers) should remain stable, as the net result of several opposing trends:

- the steady growth of transport fuel requirements, together with greater use of automotive gas oil in place of petrol;

- an increased requirement for petroleum products for non-energy uses (+1.3% per annum);

- a slight growth in the consumption of both heavy and light fuel oil by industry up to 1990 as the economy recovers, followed by a fallback in the face of competition from natural gas, steam coal and electricity;

- a steady decline in the use of gas oil for space heating in the residential and tertiary sector;

- a considerable reduction in the consumption of residual fuel oil in power stations up to 1990, followed by relative stability at 1990 levels until 2000.

Overall, these market trends will result in a change in the structure of a barrel of oil, with refining concentrated increasingly on the light and medium grades.

Structure of Community consumption of petroleum products (excluding bunkers)

	1973	1980	1983	1990	2000 %
Gaseous products	3	3	4	3	3
Petrol and naphtha	19	28	33	34	35
Gas oil and low-viscosity fuel oil	37	34	36	37	37
Residual fuel oil	37	30	22	21	19
Other	4	5	5	5	6
Total	100	100	100	100	100

The petrol and naphtha share will go up from 33% to 35% between 1983 and 2000, while the trend of heavy fuel oil will be the reverse, falling from 22% to 19% over the same period. The shares of other products will show moderate growth.

9. These trends, which were even more marked between 1973 and 1983, have already given rise to important decisions in the Community refining industry:

- closure of surplus capacity on account of the fall in consumption (regarded as structural);

- investment in conversion plants, in response to the changing structure of oil consumption.

The projections to the year 2000 continue the trends observed over the period 1973-83 and thus confirm that the right decisions were taken. They even show that further adaptation could be necessary, first to bring the products marketed into line with statutory standards (lead-free petrol, low-sulphur fuel oil) and secondly to ensure that conversion capacity is adapted to cope with structural changes in consumption. In the first case, finishing plant will need to be developed (isomerization, alkylation, production of MTBE, desulphurizing etc.) and in the second case advanced conversion plant (hydrocracking, coking, flexicoking etc.). Nevertheless, the question arises whether the additional investment earmarked for increasing conversion capacity in the Community will still be needed in order to adapt refineries to the demand structure of the year 2000, and how far imports of light and medium petroleum products refined in the oil-exporting countries of the Gulf and North Africa will meet the demand under acceptable political and economic conditions.

(b) Natural gas

10.   The predicted levels of natural gas production and contracted offtake in 1990 will meet the requirements of the consumer sectors, assuming a 1.9% annual growth in demand between 1983 and 1990. The supply conditions described above will even make adequate flexibility possible at Community level, provided that transfers can if necessary be made between Member States finding themselves with excess/inadequate supplies over a short period.

It must be emphasized that greater integration of the gas networks in the Community would not only make the transport system less wasteful by increasing its flexibility, but would improve security of supply and facilitate the development of a competitive market and of a spot market (today in its infancy).

The trend of the natural gas market after 1990 directly depends on the price assumptions in the reference projection. If natural gas prices remain tied to petroleum product prices, then gas will make further headway only on the premium markets, mainly the residential and tertiary sector. Overall, natural gas consumption will continue to grow by 0.3% a year between 1990 and 2000.

11.   Large additional quantities of natural gas - 35 to 75 million toe at today's estimate - could definitely be available for supplying the Community in 2000. If the terms of contract were economically favourable, these quantities would mean the increasing replacement of heavy fuel oil in industry and large-scale use of natural gas for the combined production of heat and power in industry instead of fuel oil and coal.

(c) Solid fuels

12. Community consumption of solid fuels is expected to increase consistently between 1983 and 1990 (+1.9% a year) in response to demand from the electricity sector and a certain return to coal in industry.

    Between 1983 and 1990, coal consumption in power stations will increase by 22 million toe and that of lignite and peat by 4 million toe. At the same time, solid fuel consumption will increase by 4 million toe in industry, but will drop back by 2 million toe in the residential and tertiary sector.

13. After 1990, demand for coal will grow much more slowly (by 0.8% per annum), returning to a rate close to that for energy demand as a whole. Power stations will again provide the principal impetus of growth, demand for coal probably growing by about 18 million toe between 1990 and 2000.

    In other consumer sectors, the assumptions are that the use of coal:

    - in industry will grow by a further 6 million toe in response mainly to the expansion of the steam-raising market;

    - will grow slightly as a raw material in the coal chemicals industry; and

    - will continue to decline in the residential and tertiary sector, although increasing quantities of distributed steam, produced in CHP stations, will be obtained from coal recorded as allocated to power stations.

(d) Electricity

14. The reference projection sets final energy demand in the year 2000 at 763 million toe, of which electricity accounts for 138 million toe. The final consumption of electricity will thus increase by 2.2% per year from 1983 to 2000, and it is expected to cover about one third of the increase in final energy demand, its share rising from 15% in 1983 to 18% in 2000. In 1973 its share was a little over 10%.

    This increased penetration of electricity on the consumer markets will result from a major expansion in specific uses of electricity, but also in some countries from the increased competitiveness of electricity in thermal uses. In the reference case, however, its penetration is limited by the general improvement in the efficiency of electrical equipment.

    The major factors of uncertainty in any forecast of electricity demand must be emphasized. The assumed rate of growth implies **some break in the link between economic growth and electricity demand.** In the past the coefficient has always been greater than 1, whereas over the period to 2000 is projected here to be only 0.85.

15. Power stations have been affected by four major trends over the last few years:

- the increasingly large proportion of base-load electricity generated by nuclear power stations;

- the return to coal for generating base-load electricity in the non-nuclear countries and for meeting middle-load nearly everywhere in the Community;

- the marked cutback in the use of heavy fuel oil in power stations, with rare exceptions; and

- the increasing restriction of use of natural gas to peak-load power stations.

These trends should continue over the next fifteen years, despite a fairly pronounced slowdown in the rate of growth of electricity generation. By the year 2000, 43% of electricity will be nuclear-generated and 39% will be generated by solid-fuel plants. Electricity from oil- or gas-fired stations will account for no more than 8% of total production.

Total net electricity production by energy source

TWh

EUR 10	1973	(%)	1980	(%)	1983	(%)	1990	(%)	2000	(%)
Coal	301.1	(30)	412.8	(34)	423.3	(34)	513	(34)	599	(32)
Lignite and peat	78.1	(8)	98.1	(8)	106.3	(9)	118	(8)	134	(7)
Petroleum products	312.2	(32)	264.8	(22)	158.5	(13)	83	(5)	71	(4)
Natural gas	100.5	(10)	107.6	(9)	95.7	(8)	101	(7)	56	(3)
Derived gas	23.8	(2)	21.1	(2)	16.1	(1)	18	(1)	16	(1)
Nuclear	53.5	(5)	149.4	(12)	275.0	(22)	534	(35)	792	(43)
Hydro & geothermal	112.9	(12)	148.8	(12)	147.4	(12)	150	(10)	165	(9)
Other	4.6	(1)	6.4	(1)	7.0	(1)	8	(1)	14	(1)
Total	986.7	(100)	1209	(100)	1229.3	(100)	1523	(100)	1847	(100)

16. Optimal use of the Community's generating capacity means that by 1990 nuclear power stations and coal fired stations should supply base load and middle load electricity, and oil- and gas-fired stations should be used for standby and peak load generation.

The following table shows the corresponding trend in energy consumption in the Community's power stations.

Energy consumed in electricity generation
million toe

EUR 10	1973	(%)	1980	(%)	1983	(%)	1990	(%)	2000	(%)
Solid fuels	101.3	(43)	130.1	(47)	133.9	(47)	160	(44)	178	(40)
Petroleum products	75.0	(32)	60.9	(22)	36.9	(13)	20	(5)	18	(4)
Gas (natural & derived)	30.6	(13)	31.3	(11)	26.2	(9)	27	(7)	15	(3)
Nuclear[5]	17.7	(7.5)	42.7	(15)	76.1	(26)	145	(40)	215	(48)
Hydro & geothermal[6]	9.1	(4)	12.3	(4)	12.0	(4)	13	(3)	14	(3)
Other	1.2	(0.5)	1.7	(1)	1.7	(1)	2	(1)	4	(1)
Total	234.2	(100)	279.0	(100)	285.8	(100)	367	(100)	444	(100)

17. The trend in the structure of electricity generation shown above assumes that major investment will be made in the industry, especially in coal-fired and nuclear plants, over the period 1983-2000. Investment should be of the order of 100 GWe in nuclear capacity and 40 GWe in solid-fuel plants. These changes will profoundly alter the structure of the Community's electricity generating capacity. Nuclear power will account for 35% of installed capacity in 2000 compared with 15% in 1983. Conventional coal-fired power stations will keep their 35% share throughout the period, the substitution process phasing out conventional plant which can operate only on hydrocarbons.

Structure of the Community's electricity generating capacity
GW, net power output

EUR 10	1983	1990	1995	2000
Hydro[7]	52.3	60.4	63.9	65.8
Nuclear	51.6[8]	94.1	119.1	144.5
Conventional with coal	120.2[8]	128.8	131.9	141.1
Conventional without coal	111.8[8]	98.7	80.5	59.6
Other[9]	0.5[8]	1.0	1.5	2.8
Total	336.4	383.1	397.0	413.8

*

*   *

---
[5] Fission heat is recorded as a primary source in accordance with the SOEC method.

[6] Hydroelectric power and electricity of geothermal origin are recorded as primary sources in accordance with the SOEC method.

[7] Including pumping stations.

[8] Estimated.

[9] Geothermal, incineraters, wind power.

CHAPTER IV

REFERENCE PROJECTION: Main features for each Member State

The reference projection set out in the previous chapters covers the European Community as a whole, but it is the result of combining the individual projections for each Community Member State.

The country projections are based on sets of assumptions which are compatible from one country to another and stem from the general socio-economic development scenario adopted for the Community.   These assumptions may differ on some points from the assumptions used by national governments, perhaps in their most recent energy projections.

For the definition of the set of economic assumptions, account was taken of specific features of each country. This applies especially as regards both volumes and prices for each Member State's energy supply and the determination of selling prices to consumers (inclusive of all taxes) of the various forms of energy, although some general simplifying assumptions were made for all Member States (Chapter I).

This chapter presents the main results obtained for each country under the reference projection.

BELGIUM

A. ECONOMIC REFERENCE FRAMEWORK

1. For the period 1983 to 2000, the GDP growth assumption is an average annual rate of 2.0%. This will be spread unevenly over the period: 1.8% from 1983 to 1990 and 2.2% from 1990 to 2000;

Between 1983 and 1990, the internal components of GDP (consumption, investment) will rise slightly and its external components (imports, exports) will show a sustained increase; their contributions to general growth will not be the same, but together they will have a positive net impact. Consumption and public investment will bear the brunt of the reorganization of public finances (down by 1.0% and 1.7% on average per annum). Private consumption will increase by only 0.7% per annum. The trend in gross fixed capital formation will differ according to its components. Expenditure on building will remain fairly low owing to the reduced level of public and household investment (+0.6% per annum on housebuilding). On the other hand, expenditure on machinery and equipment will pick up as a result of an improvement in companies' profit margins. The positive growth differential in foreign trade variables will increase during the period. On average, exports will increase at the rate of 3.3% per annum against only 1.7% for imports, which will meet the resistance of a sluggish domestic market.

The trends observed over the period 1983 to 1990 will largely be continued from 1991 to 2000. In the second period, the pattern of private and public consumption and public investment can be expected to improve, leading to slightly higher growth of GDP (up by 2.2% per annum).

At the sectoral level, the assumptions led to the following results:

- Consumer goods industries like agriculture and the food and textile industries are directly affected by budget policy to restrict personal disposable income.

- The fall in public investment combined with depressed demand in housebuilding will initially affect the building and non-metallic mineral sectors. In 1985 the level will be much lower than in 1980, but there will be a slight recovery thereafter.

- The capital goods industry will be on a sound footing throughout the period as companies' finances improve.

- Thanks to its export capacities, the chemical industry will obtain more or less satisfactory results.

- The service sector will grow at a faster rate than GDP.

On the whole, growth will be fuelled mainly by the service and capital goods sectors, which accounts for their growing shares in value added generated by the productive system (Table 1).

Table 1: Sectoral trends

	1970	AAGR[1]	1980	AAGR	1983	AAGR	'000 million BFR 1975 1990	AAGR	2000
Agriculture	63.8	0.8	69.4	1.7	72.9	-0.7	69.4	0.0	69.4
Industry	574.2	3.3	794.4	-0.8	774.4	1.8	878.2	2.0	1075.0
EII[2]	169.9	3.2	232.8	-2.1	218.4	0.6	227.4	0.9	248.6
Other industries	404.3	3.3	561.6	-0.3	556.0	2.3	650.8	2.4	826.4
Building	140.5	2.5	179.1	-7.5	141.8	0.3	145.3	2.0	177.1
Services	1037.0	3.3	1440.4	1.1	1489.9	1.7	1674.7	2.5	2143.8
Total value added (GDP)	1815.5	3.2	2483.3	-0.1	2479.0	1.8	2776.9	2.2	3465.3

B. TOTAL FINAL ENERGY DEMAND

2. Total final energy demand will show moderate growth from 1983 to 2000, after a marked fall between 1980 and 1983 caused by the low level of industrial activity in the early 1980s.

Growth up to 2000 (averaging 1.2% per annum) will be fuelled by low but dissimilar growth rates in consumption in three sectors – industry, transport, and residential and tertiary. In industry, less energy-intensive techniques and structural effects will prevent energy consumption from increasing as fast as the growth rate of industrial activity (2.0% per annum on average); in transport, the organization of traffic and improved specific vehicle consumption will maintain energy consumption at present levels; in the residential and tertiary sector, despite the stability of the housing stock (in structure and age) and intensive energy saving measures, consumption will rise relatively faster than in the other two sectors.

The structure of the final energy balances for industry and the residential and tertiary sector will not show any radical shifts markedly affecting one or other form of energy: in industry, oil products will gradually surrender a moderate share of their market to electricity, while gas and coal will keep the same shares; in the residential sector, electricity and gas will slightly increase their shares at the expense of gas oil.

----
[1] AAGR: annual average growth rate.

[2] EII: energy-intensive industries: metals, non-metallic minerals, paper, chemicals.

(a) Industry

3.   Even if the energy-intensive heavy industries grow only slowly as the
     reference projection suggests, they will still account for the
     largest share of energy consumption by industry in the year 2000.
     This will make for an average annual growth rate of 1% in total
     energy consumption by industry from 1983 to 2000, mainly concentrated
     in the period 1983-90 as the rate falls after 1990 to 0.2% per year.

     Up to 2000, heavy industry's high-temperature uses will continue to
     account for a dominant share of 60% of thermal uses as against 65% in
     1980. Production plans for energy-intensive products (steel, cement,
     basic chemicals) partly explain this inertia.

     However, structural shifts towards the capital goods industries and
     technical and energy improvements to all industrial plant will bring
     about an improvement of nearly 17% in the energy intensity of
     industry,[3] which will decline from 13.000 toe/BFR 1.000 million
     (1975) in 1983 to 11.2 in 2000, most of the fall taking place in the
     period 1990-2000.

4.   Competition between energy carriers before 1990 will be inconclusive
     until 1990 owing to the structural decline in consumption by the
     industrial sector. Oil products will temporarily supply the increase
     in total fuel demand after 1985 up to 1990-95, but gas may then take
     over. The consumption of solid fuels will be affected by the cutback
     plan for steel at the beginning of the period but will then
     stabilize; electricity consumption will continue to grow at the same
     steady pace as specific electricity requirements. The use of CHP and
     district heating will continue to expand throughout the period.

(b) Transport

5.   The level of energy consumption in the transport sector will not
     change very much between now and the year 2000 (approximately
     6 million toe); but its structure will change to some degree since
     the share of energy consumption for passenger transport in the total
     will fall slightly (52% in 2000 against 57% in 1980).

     The private car will play the same role in passenger transport in
     2000 as in 1980, for both urban transport (94% of traffic) and
     inter-city transport (88%). Improvements in specific vehicle
     consumption (-2.2 litres/100 km) will offset the increase in the
     number of passenger-kilometres between now and 2000. These
     improvements will even cause energy demand to fall by 5% in 2000
     compared with its 1983 level.

     Energy consumption for goods transport, on the other hand, will
     continue to grow, albeit more slowly (at 1.2% per annum) than
     economic activity, to which it is generally closely tied. The largest
     share of goods traffic will continue to be carried by road (65% in
     1980 and 71% in 2000).

----

[3] The energy intensity of industry is defined as the ratio of energy
    consumption (in 10^3 toe) to total value added of the industrial sector
    (in BFR 10^9 at 1975 prices).

(c) Residential and tertiary sector

6.   Energy consumption in the residential and tertiary sector fell
     significantly between 1980 and 1983, certainly as a result of greater
     practical attention to energy saving but also as a result of the
     slowdown in economic activity and wage moderation together with
     milder weather. Since the economic scenario adopted for the period
     after 1983 is more favourable than the economic environment of the
     previous three years, it seemed preferable to compare the projections
     obtained for the residential and tertiary sector with actual
     consumption in 1980.

7.   Energy consumption in the residential and tertiary sector in the year
     2000 will be only 10.7% higher than in 1980. The link with general
     economic activity will thus have been broken, as shown by the 21%
     decline in energy intensity[4] over 20 years: from 5 270 toe/BFR 1 000
     million (1975) to 4 970 in 1990 and 4 180 in the year 2000.

     There will be many mutually offsetting factors in the residential
     sector that will keep the increase in consumption between 1983 and
     2000 down to 10%:

     - ensuring relatively stable demand: the number of dwellings will
       increase only slightly (by 12%), the share of single-family houses
       will remain much the same (around 73% of the housing stock) and
       there will be only a moderate rate of house replacement (in 2000,
       only 28% of dwellings will be less than 25 years old);

     - encouraging energy saving: new insulation standards for new
       buildings and insulation of older dwellings will ensure reductions
       in the average consumption of dwellings of 5% (older dwellings) to
       20% (new dwellings); increased use of electric and gas heating
       systems with high relative outputs will ensure savings in final
       demand;

     - discouraging lower energy consumption: a marked trend towards the
       use of central heating in all types of housing (48% of heating
       systems were central systems in 1980 compared with 63% in 2000) and
       more spacious dwellings in 2000.

     The net result of these trends will be that the average useful-heat
     requirements per dwelling will be only slightly less (4%) than in
     1980, which means an increase in consumption of about 6% for the
     total housing stock from 1983 to 2000.

     The situation will be much the same for domestic hot water; and
     specific electricity requirements will be some 2% higher in 2000 than
     in 1980.

     Energy consumption in the tertiary sector will be the same in 2000 as
     it was in 1985, despite the increase in economic activity in this
     sector. This will be due to the fact that the slightly higher total
     volume of office buildings, education establishments and shops will
     be offset by better insulated buildings and more efficient heating

     _____
     [4]The energy intensity of the residential and tertiary sector is defined
     as the ratio of the total energy consumption of the sector (in 10^3 toe)
     to GDP (in BFR 10^9 at 1975 prices).

equipment.   Specific   electricity   consumption   (including air-conditioning) will rise more quickly than in the residential sector (by 3.3% per year from 1980 to 2000).

8. The most marked changes in the energy market for the whole sector will be increased shares for electricity and gas and a declining share for oil products. Between 1980 and 2000 the share of electricity will rise from 12% to 19% and that of gas from 27% to 31%, offsetting the decline in the shares of gas oil from 52% to 46% and coal from 8% to 3%.

Table 2:   Final energy consumption

	Million toe				Annual percentage changes	
	1980	1983	1990	2000	1990/83	2000/90
Industry	13.0	10.1	11.6	12.0	2.0	0.3
Transport	5.8	5.8	5.8	6.0	0.0	0.3
Residential and tertiary	13.1	10.8	13.8	14.5	3.6	0.4
Non-energy uses	2.8	2.8	3.0	3.6	1.0	1.9
Total	34.7	29.5	34.2	36.1	2.1	0.5
– solid fuels	5.4	4.0	4.4	4.0	1.4	-0.9
– oil products	16.9	14.5	17.0	17.3	2.3	0.2
– gas	8.2	6.9	8.0	8.6	2.1	0.7
– electricity	3.7	3.8	4.2	5.4	1.4	2.5
– other[5]	0.4	0.3	0.6	0.8	10.4	2.3

C. ENERGY SUPPLY

9. Table 3, which is a breakdown of gross inland primary energy consumption (GIC) based on the economic scenario adopted, shows an increase in GIC of some 2.0% per annum up to 1990 and of 1.0% per annum from 1990 onwards, making an average annual growth rate of 1.3% over the period 1983–2000.

The most important structural change is the increasing share of nuclear power, which will rise to 21.0% in 1990 and 25.2% in 2000 against 15.2% in 1983. The losers will be coal products, whose share will fall from 22.7% in 1983 to 20.0% in 1990 and then level off, and oil products, whose share will decline to 40.1% in 1990 and 37.6% in 2000 against 43.9% in 1983.

The Belgian nuclear programme will reduce dependence on imported energy from 79% in 1983 to 70% in 1990 and 67% in 2000. A closer analysis of these aggregate energy supply trends reveals movements that are much more marked.

---
[5]Heat and renewable energy sources.

Table 3:   Gross inland primary energy consumption (GIC)

				million toe
Form of energy	1983	1990	1995	2000
Coal products	9.1	9.4	9.8	10.1
Oil products	17.6	18.5	19.4	19.1
Natural gas	7.1	8.4	8.0	8.5
Nuclear fuels	6.1	9.7	11.2	12.8
Primary electricity	0.0	0.0	0.0	0.0
Renewable energy sources	0.1	0.1	0.2	0.2
GIC	40.1	46.1	48.6	50.7

(a) Oil products

10. Although the share of oil products in gross inland consumption will decline, the level of consumption will be relatively stable throughout the period. This apparent stability is the upshot of several trends:

- lower sales of heavy fuel oil to power stations, down from 3.0 million toe in 1982 to 0.2 million toe in 1990; this will of course be due to the addition of new nuclear power stations now being built to electricity generating capacity and to the conversion of a number of fuel oil/gas power stations to coal;

- a small increase in deliveries to the transport sector where the shift to diesel vehicles (13% of vehicles in 1990 and 18% in 2000) will cut petrol consumption and increase the use of diesel fuel;

- an increase in naphtha sales to the petrochemical sector, to 1.3 million toe in 1990 and 1.6 million toe in 2000 from 1.2 million toe in 1982;

- heating oil deliveries to the residential and tertiary sector will return to their 1980 level after the lower demand in 1981, 1982 and 1983, when the weather was particularly mild;

- relative stability of heavy fuel oil sales to industrial sectors; but the dual-fuel operation of many plants may mean that the changes will be rung between natural gas and industrial fuel oil as economic conditions dictate.

11. Heating oil sales remaining stable at their 1980 level and rising diesel fuel sales will increase the share of medium products in inland consumption to 50.3% in 1990 and 50.8% in 2000 against 44.6% in 1982 and 39.2% in 1980. In structural terms this change will be at the expense of heavy products and will mean that greater use will have to be made of conversion plants to satisfy inland demand.

Table 4:   Structure of consumption of oil products

Products	1980	1983	1990	1995	% 2000
Gaseous products	2.7	2.8	3.9	4.0	3.9
Light products	21.7	27.3	26.2	25.5	24.9
Medium products	39.2	40.3	50.3	49.1	50.8
Heavy products	36.4	29.6	19.6	21.5	20.4

On the basis of refining capacity at 1 January 1983, i.e. 34.6 million tonnes of atmospheric distillation capacity, 4.5 million tonnes of catalytic cracking and 2.6 million tonnes of visbreaking, the Belgian refining sector is particularly well equipped to cope with changes in the short and medium term in production structures and to ensure a utilization rate close to 75% for the whole of the oil industry, enough to guarantee its profitability. This is the result of restructuring, with the closing of two simple refineries accounting for 16.4 million tonnes of capacity, more than 30% of capacity at 1 January 1980.

(b)   Natural gas

12.   For ease of interpretation of the results, it should be noted that we have assumed the Dutch contract will run until 1996 and that the Algerian contract will be reduced to 3 000 million Nm^3 per annum after 1985 from the 5 000 million Nm^3 per annum initially agreed. This surplus situation, which is common to other Community countries, is due to the long-term energy forecasts established between 1975 and 1980 which were used as a reference basis for the negotiation of new contracts for natural gas imports to begin in 1980-90. Since then, energy consumption forecasts have been revised sharply downwards and a significant percentage of contract volumes will be surplus on the market. Corrective measures could be taken in two directions: renegotiation (as far as possible) with the exporting countries of the clauses relating to contract volumes, the assumption we have generally chosen; and an attempt to find the best way of disposing of surplus volumes on the inland market.

13.   In this general context, the pattern of Belgian natural gas consumption will follow from current trends, namely a reduction in supplies to power stations, an increase in sales to the residential and tertiary sector and level sales to industry.

14.   In contrast to medium-term trends, additional import contracts will be necessary in the year 2000 for some 3 000 million Nm^3, most of which will be supplied by the Netherlands.

A point to be noted is the very unfavourable impact which the Algerian contract has had on the average free-at-frontier price of Belgian natural gas imports and, consequently, on the competitiveness of natural gas on its main markets. If the economic clauses of this contract were revised downwards, imports of natural gas could undoubtedly increase appreciably from 1990 onwards in view of the potential outlets, primarily on industrial markets.

(c)   Coal products

15.   Although the prices of coal products are attractive, consumption will remainevel in the medium term and their share of gross inland consumption will fall to 20% against 24.2% in the 1980. This trend is due to falling orders from the two main consumers: the electricity sector, which is gradually replacing its basic plant (now partly fuelled by coal) by nuclear power stations, and the iron and steel industry, which is introducing technical improvements to reduce coke consumption. This decline will be offset by the steady shift of industrial sectors to coal, but sales to households will continue to fall.

After 1990 the consumption of solid fuels in GIC will rise slightly and their share will hold steady, owing to a substantial increase in sales for electricity generation. It should also be noted that for combined heat/electricity production coal will gradually become the economically ideal fuel despite the increased investment which it necessitates.

As regards the exploitation of indigenous resources, the medium-term trend is towards an output of 5 million t of coal, given the high production costs stemming from the extremely unfavourable operating conditions. However, in the long term underground gasification could be expected to make a modest breakthrough, producing a gas of low calorific value for use in power stations. This will mark the success of one of the main objectives of current R&D policy in Belgium.

(d)   Electricity sector

16.   The basic feature of the electricity sector is the increasing role played by nuclear power. Demand will grow at an annual rate of 1.3% between 1983 and 1990 and 2.6% between 1991 and 2000, i.e. an average annual growth rate of 2.1% over the whole period.

The capital investment planned for the period 1980-90 in the centralized sector will be generally sufficient to meet the increase in demand. Over the period 1990-2000 nuclear electricity capacity will continue to grow with the installation of 2047 MW (participation in the French power stations at Chooz and further installed capacity of 1700 MW). Other investments will be required for continuation of the coal conversion programme for dual fuel (oil/gas) power stations and installation of peak load plants.

Table 5:   Capital investment programme for the electricity sector

Units	1980–1990		1991–1995		1996–2000	
Nuclear PWR	3765	MWe	857	MWe	1190	MWe
Conventional plants	38	MWe	23	MWe	214	MWe
Conversion to coal	815	MWe	510	MWe	225	MWe
CHP	50	MWth	240	MWth	185	MWth

The overall breakdown of electricity production (all producers) by
type of fuel shows the increasing share taken by nuclear power, which
from 1990 will level off at 70% of electricity produced;   this fits
in with the technical constraint of non-modulation introduced into
the model.   It is obviously liquid and gaseous fuels which will feel
the effects of competition from nuclear power,   since their use will
be increasingly confined to peak load generation,   whereas coal,
having been displaced for base load generation by nuclear power, will
find its role as the fuel for central load stations.

Table 6:   Total net electricity production (in TWh and %)

	1983		1990		1995		2000	
	TWh	%	TWh	%	TWh	%	TWh	%
Hydro	1.2	2.3	0.4	0.8	0.4	0.7	0.4	0.6
Nuclear	22.8	45.7	36.5	68.6	42.2	70.0	48.0	70.0
Solid fuels	13.7	27.5	11.7	22.0	13.0	21.6	13.7	20.0
Liquid fuels	16.4	13.0	0.5	0.9	0.4	0.7	2.0	2.9
Gaseous fuels	5.5	11.1	3.8	7.1	3.8	6.3	4.0	5.8
Other	0.2	0.4	0.3	0.6	0.4	0.7	0.5	0.7
Total	49.9	100.0	53.2	100.0	60.2	100.0	68.6	100.0

D. SUMMARY TABLES

ENERGY SUPPLY/DEMAND BALANCE   1980   **BELGIUM**

million toe

	SOLID FUEL	PETROL PRODCT	GAS	NUCLR ENERGY	HEAT	HYDRO+ OTHERS	RENEW ENER	TOTAL
PRIM PRODUCT	4.7	-	-	3.1	-	0.0	-	7.8
TOT IMPORTS	7.9	43.4	9.0	-	-	0.6	-	60.8
TOT EXPORTS	0.9	17.7	0.0	-	-	0.8	-	19.4
STOCK CHANGE	-0.6	-0.4	-	-	-	-	-	-1.0
GROSS CONSUM	11.1	25.2	9.0	3.1	-	-0.2	-	48.2
BUNKERS	-	2.4	-	-	-	-	-	2.4
INLAND CONSM	11.1	22.8	9.0	3.1	-	-0.2	-	45.8
ELEC POWER S	3.0	4.0	2.4	3.1	-0.4	-4.5	-	7.6
OTHER TRANSF	2.7	0.3	-2.3	-	-	-	-	0.1
ENERG INDUST	0.0	1.8	0.7	-	-	0.6	-	3.1
FINAL CONSUM	5.4	16.9	8.2	-	0.4	3.7	-	34.7
- NON ENERGY	0.2	2.1	0.6	-	-	-	-	2.8
- INDUSTRY	4.1	2.4	4.0	-	0.4	2.1	-	13.0
- TRANSPORT	-	5.7	-	-	-	0.1	-	5.8
- HOUSEHOLD	1.1	6.8	3.6	-	0.0	1.6	-	13.1
STAT DIFFER	0.0	-0.2	-0.0	-	-	-	-	-0.2

ENERGY SUPPLY / DEMAND BALANCE - 1983

million toe

	SOLID FUEL	PETROL PRODCT	GAS	NUCLR ENERGY	HEAT	HYDRO+ OTHERS	RENEW ENER	TOTAL
PRIM PRODUCT	4.5	-	0.0	6.1	-	0.0	0.1	10.8
TOT IMPORTS	5.4	36.0	7.2	-	-	0.4	-	49.0
TOT EXPORTS	1.0	16.1	-	-	-	0.4	-	17.5
STOCK CHANGE	0.2	0.2	-0.1	-	-	-	-	0.3
GROSS CONSUM	9.1	20.1	7.1	6.1	-	-	0.1	42.6
BUNKERS	-	2.5	-	-	-	-	-	2.5
INLAND CONSM	9.1	17.6	7.1	6.1	-	-	0.1	40.1
ELEC POWER S	3.4	1.6	1.5	6.1	-0.3	-4.4	0.1	8.0
OTHER TRANSF	1.8	0.2	-1.9	-	-	-	-	0.1
ENERG INDUST	-	1.3	0.6	-	-	0.6	-	2.6
FINAL CONSUM	3.9	14.5	6.9	-	0.3	3.8	-	29.4
- NON ENERGY	0.2	2.1	0.5	-	-	-	-	2.8
- INDUSTRY	2.8	2.2	2.8	-	0.2	2.0	-	10.0
- TRANSPORT	-	5.7	-	-	-	0.1	-	5.8
- HOUSEHOLD	0.9	4.5	3.6	-	0.1	1.7	-	10.8

ENERGY SUPPLY/DEMAND BALANCE   1990   **BELGIUM**
                                                              million toe

	SOLID FUEL	PETROL PRODCT	GAS	NUCLR ENERGY	HEAT	HYDRO+ OTHERS	RENEW ENER	TOTAL
PRIM PRODUCT	4.2	–	0.0	9.7	–	0.0	0.1	14.0
TOT IMPORTS	5.7	29.6	8.8	–	–	0.4	–	44.5
TOT EXPORTS	0.5	8.7	0.4	–	–	0.4	–	10.0
STOCK CHANGE	–	–	–	–	–	–	–	–
GROSS CONSUM	9.4	20.9	8.4	9.7	–	0.0	0.1	48.5
BUNKERS	–	2.4	–	–	–	–	–	2.4
INLAND CONSM	9.4	18.5	8.4	9.7	–	0.0	0.1	46.1
ELEC POWER S	2.8	0.2	1.0	9.7	-0.6	-4.9	0.1	8.3
OTHER TRANSF	2.2	0.2	-1.5	–	–	–	–	0.9
ENERG INDUST	–	1.1	0.9	–	–	0.7	–	2.7
FINAL CONSUM	4.4	17.0	8.0	–	0.6	4.2	0.0	34.2
– NON ENERGY	0.2	2.0	0.8	–	–	–	–	3.0
– INDUSTRY	3.4	2.5	3.0	–	0.5	2.2	–	11.6
– TRANSPORT	–	5.7	–	–	–	0.1	–	5.8
– HOUSEHOLD	0.8	6.8	4.2	–	0.1	1.9	0.0	13.8

ENERGY SUPPLY/DEMAND BALANCE   1995
                                                              million toe

	SOLID FUEL	PETROL PRODCT	GAS	NUCLR ENERGY	HEAT	HYDRO+ OTHERS	RENEW ENER	TOTAL
PRIM PRODUCT	4.2	–	0.0	11.2	–	0.0	0.2	15.6
TOT IMPORTS	6.1	30.5	8.4	–	–	0.4	–	45.4
TOT EXPORTS	0.5	8.7	0.4	–	–	0.4	–	10.0
STOCK CHANGE	–	–	–	–	–	–	–	–
GROSS CONSUM	9.8	21.8	8.0	11.2	–	0.0	0.2	51.0
BUNKERS	–	2.4	–	–	–	–	–	2.4
INLAND CONSM	9.8	19.4	8.0	11.2	–	0.0	0.2	48.6
ELEC POWER S	3.3	0.2	0.8	11.2	-0.7	-5.6	0.2	9.4
OTHER TRANSF	2.3	0.2	-1.5	–	–	–	–	1.0
ENERG INDUST	–	1.2	0.8	–	–	0.8	–	2.8
FINAL CONSUM	4.2	17.8	7.9	–	0.7	4.8	0.0	35.4
– NON ENERGY	0.2	2.3	0.9	–	–	–	–	3.4
– INDUSTRY	3.4	3.0	2.6	–	0.6	2.4	–	12.0
– TRANSPORT	–	5.8	–	–	–	0.1	–	5.9
– HOUSEHOLD	0.6	6.7	4.4	–	0.1	2.3	0.0	14.1

ENERGY SUPPLY/DEMAND BALANCE   2000   **BELGIUM**

million toe

	SOLID FUEL	PETROL PRODCT	GAS	NUCLR ENERGY	HEAT	HYDRO+ OTHERS	RENEW ENER	TOTAL
PRIM PRODUCT	3.9	–	0.0	12.8	–	0.0	0.2	16.9
TOT IMPORTS	6.7	30.2	8.9	–	–	0.4	–	46.2
TOT EXPORTS	0.5	8.7	0.4	–	–	0.4	–	10.0
STOCK CHANGE	–	–	–	–	–	–	–	–
GROSS CONSUM	10.1	21.5	8.5	12.8	–	0.0	0.2	53.1
BUNKERS	–	2.4	–	–	–	–	–	2.4
INLAND CONSM	10.1	19.1	8.5	12.8	–	0.0	0.2	50.7
ELEC POWER S	3.7	0.5	0.7	12.8	-0.8	-6.3	0.2	10.8
OTHER TRANSF	2.4	0.2	-1.5	–	–	–	–	1.1
ENERG INDUST	–	1.1	0.7	–	–	0.9	–	2.7
FINAL CONSUM	4.0	17.3	8.6	–	0.8	5.4	0.0	36.1
– NON ENERGY	0.2	2.4	1.0	–	–	–	–	3.6
– INDUSTRY	3.4	2.4	2.9	–	0.7	2.6	–	12.0
– TRANSPORT	–	5.9	–	–	–	0.1	–	6.0
– HOUSEHOLD	0.4	6.6	4.7	–	0.1	2.7	0.0	14.5

Summarized Energy Balance - BELGIUM

in million toe	1973[a]	1980[a]	1983[a]	1990[b]	1995[b]	2000[b]
Gross Energy Consumption	48.81	48.12	42.62	48.50	51.00	53.10
- Bunkers	3.03	2.38	2.49	2.40	2.40	2.40
- Inland consumption	45.78	45.74	40.13	46.10	48.60	50.70
Inland Energy Consumption	45.78	45.74	40.13	46.10	48.60	50.70
- Solid fuels	11.31	10.97	9.16	9.40	9.80	10.10
- Oil	27.29	22.89	17.64	18.50	19.40	19.10
- Gas	7.19	8.91	7.12	8.40	8.00	8.50
- Primary Electricity, etc.	-0.01	2.97	6.21	9.80	11.40	13.00
.Indigenous Production[1]	5.89	7.92	10.80	14.00	15.60	16.90
- Hard coal	5.79	4.69	4.54	4.20	4.20	3.90
- Lignite & Peat	-	-	-	-	-	-
- Oil	-	-	-	-	-	-
- Natural gas	0.04	0.03	0.02	-	-	-
- Nuclear energy	0.02	3.12	6.12	9.70	11.20	12.80
- Hydro & geothermal[2]	0.01	0.03	0.03	0.00	0.00	0.00
- Others & renewables	0.03	0.05	0.09	0.10	0.20	0.20
Net Imports[3]	42.92	41.25	31.53	34.50	35.40	36.20
- Solid fuels	5.31	6.91	4.43	5.20	5.60	6.20
- Oil	30.53	25.68	19.95	20.90	21.80	21.50
- Natural gas[2]	7.14	8.89	7.18	8.40	8.00	8.50
- Electricity[2]	-0.06	-0.23	-0.03	0.00	0.00	0.00
Stock changes[4]	-0.00	1.05	-0.29	-	-	-
- Solid fuels	-0.21	0.63	-0.18	-	-	-
- Oil	0.22	0.41	-0.18	-	-	-
- Gas	-0.01	0.01	0.07	-	-	-
Electricity Generation Input	9.66	12.57	12.71	13.80	15.70	17.90
- Solid fuels[5]	2.24	3.71	3.90	3.20	3.70	4.05
- Oil	5.07	4.07	1.65	0.20	0.20	0.50
- Natural gas	2.29	1.59	0.92	0.60	0.40	0.35
- Nuclear energy	0.02	3.12	6.12	9.70	11.20	12.80
- Hydro & geothermal[2]	0.01	0.03	0.03	0.00	0.00	0.00
- Others & renewables	0.03	0.05	0.09	0.10	0.20	0.20

indicators (related to long term objectives)

	1973 -1963	1980 -1975	1983 - 1980	1990 -1983	2000 -1990
land Energy annual growth rate	+5.0%	+1.9%	-4.2%	+2.0%	+1.0%
DP annual growth rates	+4.9%	+3.0%	+0.1%	+1.8%	+2.3%
nergy-GDP ratio	1.02	0.63	-	1.11	0.43

	1973	1980	1983	1990	1995	2000
hare of oil in gross energy onsumption	62.1%	52.5%	47.2%	43.1%	42.7%	40.5%
hare of coal and nuclear in n electricity production	23.4%	54.3%	78.8%	93.5%	94.9%	94.1%
upply dependance on imports	87.9%	85.7%	74.0%	71.1%	69.4%	68.2%

Sources: a. Statistical Office of the European Communities
b. Energy 2000
Notes: 1. Production of primary sources, including recovered products.
2. The conversion of electricity, including hydro and geothermal, is based on its actual energy content: 3600 kjoules/kWh or 860 kcal/kWh
3. The (-) sign means net exports
4. The (-) sign means a stock decrease
5. Including coke oven gas and blast furnace gas (derived from coal)

D E N M A R K

A.   ECONOMIC REFERENCE FRAMEWORK

1. For the period 1983 to 2000, the GDP growth assumption is an average annual rate of 2.4%. Growth will be uneven over the period: 2.1% from 1983 to 1990 and 2.6% from 1991 to 2000.

From 1983 to 1990, public consumption is expected to remain practically constant, while the rate of increase of public-sector investment declines by 0.9% per annum. But a real increase in earnings should stimulate private consumption, which should grow by an average of 1.5% per annum. GDP growth will be sustained by the 4.9% per annum increase in gross fixed capital formation. As the international situation improves, exports should grow strongly (up by 3.5% per annum).

The trends observed over the period 1983-90 will broadly continue from 1991 to 2000, making for stronger GDP growth.

Industry by industry, the assumptions adopted point to the following results:

- relatively strong growth in the capital goods and chemical industries fuelled by rising investment and exports;

- vigorous growth in the food and textile industries thanks to buoyant private consumption;

- a less promising outlook for the building and non-metallic minerals sectors, which will be helped by productive investments but hindered by the stagnation in housebuilding and in public investment;

- services will continue to grow at the same rate as GDP.

These trends are given in detail in Table 1.

Table 1:   Sectoral trends (value added by sector)

Unit: 1000m DKR '75	1970	AAGR[6]	1980	AAGR	1983	AAGR	1990	AAGR	2000
Agriculture	9.1	2.3	11.4	3.1	12.5	0.0	12.5	0.7	13.4
Industry	40.2	2.3	50.3	0.3	50.8	2.8	61.7	2.5	79.2
EEI	10.9	1.8	13.0	-2.1	12.2	3.6	15.6	2.0	19.1
Other industries	29.3	2.4	37.3	1.1	38.6	2.6	46.1	2.7	60.1
Building	21.2	-3.5	14.8	-7.3	11.8	4.3	15.8	2.0	19.2
Services	101.8	3.2	139.7	3.1	153.3	1.9	175.0	2.8	230.7
Total value added (GDP)	172.3	2.3	216.2	1.8	228.4	2.1	265.0	2.6	342.5

B.   TOTAL FINAL ENERGY DEMAND

2.   In Denmark more than in any other country, total final energy demand
largely hinges on the residential and tertiary sector, which
accounted for approximately 55% of Denmark's energy consumption in
1983 against 21% for industry and 24% for transport.

Substantial energy savings in this key sector will allow no more than
weak growth in total energy demand, which will settle at an average
annual growth rate of 1.1% between now and the year 2000. Set against
the 2.4% annual growth rate for GDP, this reveals a clear break in
the link between energy demand and economic growth, helped by the
fact that the residential sector, where energy consumption is not
closely linked to economic activity, plays such a leading part.

Energy demand in the residential and tertiary sector in 2000 is
expected to be around the 1983 level, with little change in the
housing stock, many more dwellings insulated and heated less
wastefully, and penetration by more efficient energy vectors
(district heating and gas).

In industry energy consumption will grow strongly, by 3.6% a year
over the period 1983-2000,[8] a rate close to the growth of economic
activity in this sector because no major structural changes will have
been made: there are even plans for a slight increase in output of
products with a high energy content, which in turn will add to the
energy intensity of the sector as a whole.

Consumption in the transport sector will rise more modestly (up 0.9%
per year on average from 1983 to 2000), helped by the continued
switch from private to public transport.

3.   Before going any further, it should be noted that in Denmark final
energy consumption fell by more than 12% from 1980 to 1983, with a
drop of 27% in industry and 14% in the residential and tertiary
sector. This decline was partly due to a more rational use of energy,

---

[6] AAGR: average annual growth rate.
[7] EII: energy-intensive industries (metals, non-metallic minerals, paper, chemicals).
[8] 1983 was a year of very low energy consumption; calculated from 1980, the average annual growth rate would be 1.4%.

but the main reason was the slowdown in economic activity and the erosion of incomes. Since the scenario for the period after 1983 assumes an improvement in the economic environment, it is quite normal that from 1983 to 1990 final energy consumption should substantially increase, mainly in industry. Overall, by 1990 final energy consumption will have returned to 1980 levels.

Analysis of final consumption by form of energy shows that district heating and gas are the vectors whose share in the final energy balance will rise the fastest between 1983 and 2000, at the expense of oil products and heating oil. Electricity demand will rise at the same rate (2.0% per year) as specific requirements in the residential-a and tertiary sector and industry.

a)   Industry

4.   The fairly vigorous growth of industrial activity at roughly the same rate as economic activity in general, 2.6% a year is bound to stimulate substantial growth in energy consumption (3.6% a year on average from 1983 to 2000). Nevertheless, the energy intensity of the industrial sector will decrease by 14% in 20 years, from 60 (10^3 toe/10^6 DKR (1975)) in 1980 to 57 in 1990 and 51 in 2000.

That there will be no major restructuring of industry is the main reason for the continuing link between economic activity and energy demand: heavy industries, including the chemical industry, will account for much the same share of total value added throughout the period, with the production of energy-intensive products planned to increase at the same rate as activity in the industry. Energy productivity gains will stem solely from technical improvements in processing methods and equipment, and from improved overall energy management. The moves towards methods based on recycling (such as the recycling of scrap for steel production) and more rational use of energy should make for energy gains of, for example, 15% per tonne of steel produced and 20% per unit of value added in light industry.

5.   Greater use is likely to be made of coal and electricity in certain heat-raising applications (up to approximately 20% and 10% respectively of the market in 2000). Gas began to make its mark on the energy market in 1984; in the medium term it will successsfully penetrate industry, where it will cover up to 20% of total energy demand. These advances will reduce the share taken by oil products.

b)   Transport

6.   At present passenger and freight transport each account for roughly the same share of energy consumption in this sector, but are expected to move in opposite directions, with consumption on the passenger side decreasing between now and 2000 and consumption on the goods side growing at almost the same rate as industrial activity, to which it is closely linked.

The predominant trend in passenger transport is a distinct shift from private cars to public transport, which in 1983 already carried approximately 17% of all passenger traffic. By 2000 this share will have increased to 25%, with inter-city services playing an important role: in this category alone public transport services will increase their share from 16% to 24% in 2000. Public transport's share of urban services is expected to rise from 22% in 1983 to 28% in 2000.

This shift of traffic plus improved vehicle performance will reduce specific consumption by about 20% and cut energy consumption by passenger transport by 10% between now and 2000.

7. In contrast, energy consumption on the goods side will grow steadily (by 2.5% a year on average) over the next 17 years. Approximately 85% of the traffic will be carried by road.

c) Residential and tertiary sector

8. With so little change in the structure of the housing stock in either age or type (houses or apartment blocks), plus more widespread insulation and greater penetration by district heating, consumption in this sector will decrease slightly between 1980 and the year 2000, despite a 12% increase in the number of households from 1983 to 2000 and the buoyancy of the tertiary sector.

The energy intensity[9] of this sector will clearly reflect this trend, falling from 39 (10^3 toe/10^9 DKR (1975)) in 1980 to 29 in 1990 and to 22 in 2000, a 43% drop in 20 years.

9. The only change in the breakdown of the housing stock between single-family houses and apartment blocks will be a slight increase in the proportion of houses. The scope for change is limited by the fact that the housing stock in 1980 was relatively new and will almost all still be in use in 2000. In 2000, housing built after 1980 will account for only about 16% of the total stock.

Major savings (25% reduction in heat losses) by better insulation of pre-1980 housing, by far the majority of the stock in 2000, will reduce the average home's heating requirements by approximately 16% by 2000. The increase in the shares taken by district heating and gas will further improve energy performance since both are very efficient means of converting energy.

Specific electricity demand will grow by an average of 2.7% a year, while total energy requirements for domestic hot water will remain much the same.

All in all, these very limited changes will mean little movement in energy demand in the residential sector from 1983 to 2000.

10. Although economic activity in the tertiary sector will run slightly ahead of GDP growth with an average annual growth rate of 2.8%, this will not generate any appreciable increase in the number of office jobs or in the total area of retail trade premises. The moderate growth in these two components, which account for such a large proportion of energy consumption in the tertiary sector, is reflected

---

[9] Defined as the ratio of energy consumption in the residential and tertiary sector (in 10^3 toe) to GDP (in 10^9 DKR at 1975 prices).

in the heat requirements in this sector. Technical improvements in equipment and buildings will even lead to a fall in energy consumption in the tertiary sector (down 10% from 1980 to 2000).

11.   A breakdown of energy consumption over the whole sector shows that district heating and gas will appreciably increase their share of the heating market not only in new but also in old buildings, where they will replace heating oil. From approximately 12% of consumption in the residential sector in 1983, district heating will push its share up to 17% by 2000, and gas will show a very substantial increase, from 1% in 1983 to 17% in 2000.

Table 2: Final energy consumption

	Million Toe				Annual % changes	
DENMARK	1980	1983[10]	1990	2000	1990/83	2000/90
Industry	3.0	2.2	3.5	4.0	+6.8[11]	+1.3
Transport	3.1	3.2	3.3	3.7	+0.4	+1.2
Residential and tertiary	8.4	7.2	7.8	7.6	+1.1	-0.3
Non-energy uses	0.4	0.5	0.4	0.5	-3.2	+2.3
Total	14.9	13.1	15.0	15.8	+2.0	+0.5
of which: Solid fuels	0.6	0.5	1.2	1.6	+13.3	+2.9
- Oil products	11.6	9.6	8.6	7.8	-1.6	-1.0
- Gas	0.1	0.1	1.7	1.9	+49.9	+1.1
- Electricity	1.9	2.0	2.2	2.8	+1.4	+2.4
- Other[12]	0.7	0.9	1.3	1.7	+5.4	+2.7

C.   ENERGY SUPPLY

12.   Table 3, a breakdown of gross inland primary energy consumption (GIC) based on the economic scenario adopted, shows an increase in GIC of about 3.3% per annum up to 1990 and 0.8% per annum thereafter, making an average annual growth rate of 1.7% for the period 1983-2000.

The most important structural feature is the decreasing share of oil products, which will fall to 49% in 1990 and 43% in 2000 against 64% in 1983. Natural gas, a newcomer in the energy supply pattern, will gain from this shift: from 1985 on, its share of GIC will settle at around 8.6%. Coal products will gradually increase their share to 47% in 2000, unless a nuclear power station building programme is undertaken.

The development of Denmark's oil and gas resources will reduce dependence on imported energy from 85.6% in 1983 to 68.6% in 1990 and 69.1% in 2000. A source-by-source analysis of these aggregate energy supply trends reveals movements that are much more marked.

---

[10] Consumption by industry in 1979: 3.3 million toe.
[11] The increase over the period 1980-90 will be 1.4% per annum.
[12] Heat and renewable energy sources.

Table 3: Gross inland primary energy consumption

				million toe
Form of energy	1983	1990	1995	2000
Coal products	5.5	8.4	9.0	10.2
Oil products	10.4	10.0	9.9	9.4
Natural gas	-	1.7	1.8	1.9
Nuclear fuels	-	-	-	-
Primary electricity	0.4	0.1	0.1	0.1
Renewable energy sources	-	0.1	0.2	0.3
GIC	16.2	20.3	21.0	21.9

(a) Oil products

13. Since 1973, when oil products covered 95% of gross inland consumption, the Danish Government has been trying to reduce oil consumption by encouraging all possible substitution and conservation measures. This trend can be expected to continue until the end of the period.

Of course, this reduction in consumption of oil products, which will slow down over time, will not affect all products in the same way, on the contrary:

- substantially reduced of heating oil sales to the residential and tertiary sector as it makes a major shift to natural gas and distinct heating for its heating requirements. Only a very small part of this reduction will be made up for by the gradual conversion in the transport sector to diesel;

- steady heavy fuel oil sales to those sectors of industry whose energy consumption is growing fast;

- level sales of light products, with the increase in the offtake of kerosene compensating for the reduction in motor fuel sales.

With the stable trend in sales of light products (petrol, kerosene and naphtha) and of heavy products, the lower offtakes of heating oil will reduce its share of consumption from 46.4% in 1983 to 40.7% in 2000, giving a proportionate increase in the shares of other products.

Table 4: Structure of consumption of oil products

					%
Products	1980	1983	1990	1995	2000
Gaseous products	1.9	2.6	1.4	1.4	1.5
Light products	17.4	22.5	23.0	23.2	24.4
Medium products	43.4	46.4	47.8	43.1	40.7
Heavy products	37.3	28.5	27.8	32.3	33.4

14. On the basis of the refining capacity at 1 January 1983 (10.7 million tonnes of topping capacity, 2 million tonnes of thermal cracking plus 1.8 million tonnes of visbreaking), Denmark will still have to invest in nearly 2 million tonnes of catalytic cracking capacity by 1990 if

it wishes to reduce its imports of light and medium products. Without this, rationalization of the existing refining capacity will be unavoidable if the plant maintained in service is to remain profitable.

Finally, the Danish sections of the North Sea fields will gradually come on stream and reduce Denmark's dependence on imported oil to 62% in 1990 and 57% by 2000, against 79% in 1983.

(b) Natural gas

15.   Natural gas makes its first significant contribution to Denmark's energy supply in the decade to 1990, once the natural gas fields in the Danish section of the North Sea start producing. Denmark received the first natural gas from the Tyra field in 1984.

Output is expected to total 1.0 million toe in 1985, rising to 2.4 million toe in 1990 and then holding steady at this level. Allowing for exports, natural gas will hold an 8.4% share of gross inland consumption in 1990 and 8.7% in 2000. This contribution by gas to the energy supply will further diversify energy sources and reduce dependence on imports as the expansion comes from indigenous resources.

In addition, Denmark has no plans for long-term import contracts to supplement domestic output. Only one contract of very short duration has been negotiated – with the Federal Republic of Germany – to ensure that the transmission and distribution grid systems start up in time to enable natural gas to penetrate its potential markets and to guarantee at least a minimum market for the production from Tyra from 1984 onwards. Export contracts, in fact, with Sweden and Germany, are planned for the medium term.

16.   The results show that in the medium and long term the residential sector should absorb most of the natural gas produced. Up to 1991, however, 2 500 million Nm^3 of natural gas will be supplied under contract to power stations, to allow time for the transmission and distribution grids to be fully developed. The extension of district heating networks opens up large markets for natural gas in the large boilers used for centralized heat production, which obviates the need to build both heat and natural gas distribution networks at the same time. In addition, the prospects of further penetration by natural gas in the industrial sector also seem promising, although less firm.

(c) Coal products

17.   Since coal products are attractively priced on international markets, consumption can be expected to rise substantially over the period, from 5.5 million toe in 1983 to 8.4 million toe in 1990 and 10.2 million toe in 2000. Although the electricity sector remains the largest coal consumer and accounts for most of the increase in consumption, solid fuels will also make a substantial advance in supplying the central boilers for district heating networks and will achieve a smaller penetration in supply to industry.

The increase in supplies of solid fuel to the electricity sector between 1995 and 2000 implies that Denmark will not be using nuclear energy before the end of the century.   If the Danish Parliament, possibly after a national referendum, authorizes the development of nuclear energy, then the choice will have to be made after 1995 between coal and nuclear for installing 1200 MW.   If nuclear energy is used, this would displace 1.9 million toe of coal by 2000 and reduce the share of solid fuels to 38% of gross inland consumption.

(d) Electricity sector

The main feature in the electricity sector is the dominant share of solid fuels in meeting demand, which is expected to grow by 2.8% per annum between 1983 and 1990 and by 2.0% between 1991 and 2000, an average annual growth rate of 2.3% per annum over the whole period.

As regards investments, after the extensive programme to convert existing power stations to coal, the main emphasis up to 1995 will be on building approximately 1790 MWe of dual-fired (coal/oil) power stations combining, wherever possible, electricity production with supplying heat to district heating networks.   After 1995, as already said, a choice will be made between this type of power station and PWR stations to generate an extra 1200 MW.

Table 5: Capital investment programme for the electricity sector

Plant	1981-1990	1991-1995	1996-2000
Nuclear	-	-	-
Dual-fired	1 377 MWe	-	1 128 MWe
Renewable	5 MWe	16 MWe	45 MWe
CHP	799 MWth	300 MWth	346 MWth

The overall breakdown of electricity production (all producers) by type of fuel underlines the leading role played by solid fuels up to 2000.

Electricity imports provided nearly 10% of the electricity consumed in Denmark in 1982, but are expected to disappear almost completely in the medium term before recovering in the long run to cover about 4% of Denmark's electricity consumption in 2000.

Table 6: Total net electricity production

							TWh and %	
Origin	1983		1990		1995		2000	
	TWh	%	TWh	%	TWh	%	TWh	%
Solid fuels	19.55	94.3	26.65	90.8	28.13	89.5	32.00	91.1
Liquid fuels	1.13	5.4	2.68	19.1	3.24	10.3	2.88	8.2
Nuclear power	-	-	-	-	-	-	-	-
Renewable sources	0.06	0.3	0.02	0.1	0.08	0.2	0.23	0.7
Total	20.74	100.0	29.35	100.0	31.45	100.0	35.11	100.0

D. SUMMARY TABLES

Energy Supply / Demand Balance - 1980   **DENMARK**

million toe

	SOLID FUEL	PETROL PRODCT	NAT. GAS	NUCLR ENERGY	HEAT	HYDRO+ OTHERS	RENEW ENER	TOTAL
PRIM PRODUCT	-	0.3	0.0	-	-	-	-	0.3
TOT IMPORTS	6.1	14.9	-	-	-	0.2	-	21.2
TOT EXPORTS	-	1.7	-	-	-	0.1	-	1.8
STOCK CHANGE	-0.2	+0.2	-	-	-	-	-	0.0
GROSS CONSUM	5.9	13.7	-	-	-	0.1	-	19.7
BUNKERS	-	0.4	-	-	-	-	-	0.4
INLAND CONSM	5.9	13.3	-	-	-	0.1	-	19.3
ELEC POWER S	5.1	1.2	-	-	-0.7	-2.2	-	3.4
OTHER TRANSF	0.1	0.1	-0.1	-	-	-	-	0.1
ENERG INDUST	-	0.3	-	-	-	0.4	-	0.7
FINAL CONSUM	0.6	11.6	0.1	-	0.7	1.9	-	14.9
- NON ENERGY	-	0.4	-	-	-	-	-	0.4
- INDUSTRY	0.5	2.0	-	-	-	0.5	-	3.0
- TRANSPORT	-	3.1	-	-	-	0.0	-	3.1
- HOUSEHOLD	0.1	6.1	0.1	-	0.7	1.4	-	8.4
STAT DIFFER	0.1	0.1	-	-	-	-	-	0.2

ENERGY SUPPLY / DEMAND BALANCE - 1983 **DENMARK**

million toe

	SOLID FUEL	PETROL PRODCT	NAT. GAS	NUCLR ENERGY	HEAT	HYDRO+ OTHERS	RENEW ENER	TOTAL
PRIM PRODUCT	-	2.2	-	-	-	-	-	2.2
TOT IMPORTS	5.4	11.5	-	-	-	0.7	-	17.5
TOT EXPORTS	-	3.0	-	-	-	0.3	-	3.3
STOCK CHANGE	0.1	0.1	-	-	-	-	-	0.2
GROSS CONSUM	5.5	10.8	-	-	-	0.4	-	16.6
BUNKERS	-	0.4	-	-	-	-	-	0.4
INLAND CONSM	5.5	10.4	-	-	-	0.4	-	16.2
ELEC POWER S	5.2	0.3	-	-	-0.9	-1.9	-	2.6
OTHER TRANSF	-	0.2	-0.1	-	-	-	-	0.1
ENERG INDUST	-	0.3	0.0	-	-	0.3	-	0.6
FINAL CONSUM	0.3	9.6	0.1	-	0.9	2.0	-	12.9
- NON ENERGY	-	0.5	-	-	-	-	-	0.5
- INDUSTRY	0.3	1.3	0.0	-	-	0.6	-	2.2
- TRANSPORT	-	3.2	-	-	-	-	-	3.2
- HOUSEHOLD	0.2	4.6	0.1	-	0.9	1.4	-	7.2
STAT. DIFFER.	-0.2	-	-	-	-	-	-	-0.2

ENERGY SUPPLY / DEMAND BALANCE - 1990

million toe

	SOLID FUEL	PETROL PRODCT	NAT. GAS	NUCLR ENERGY	HEAT	HYDRO+ OTHERS	RENEW ENER	TOTAL
PRIM PRODUCT	-	4.0	2.4	-	-	-	0.1	6.5
TOT IMPORTS	8.4	7.8	-	-	-	0.1	-	16.3
TOT EXPORTS	-	1.4	0.7	-	-	-	-	2.1
STOCK CHANGE	-	-	-	-	-	-	-	-
GROSS CONSUM	8.4	10.4	1.7	-	-	0.1	0.1	20.7
BUNKERS	-	0.4	-	-	-	-	-	0.4
INLAND CONSM	8.4	10.0	1.7	-	-	0.1	0.1	20.3
ELEC POWER S	7.2	0.7	-	-	-1.4	-2.4	0.0	4.1
OTHER TRANSF	-	-	-	-	-	-	-	-
ENERG INDUST	-	0.7	0.0	-	0.2	0.3	-	1.2
FINAL CONSUM	1.2	8.6	1.7	-	1.2	2.2	0.1	15.0
- NON ENERGY	-	0.4	-	-	-	-	-	0.4
- INDUSTRY	0.6	1.5	0.6	-	0.2	0.6	-	3.5
- TRANSPORT	-	3.3	-	-	-	0.0	-	3.3
- HOUSEHOLD	0.6	3.4	1.1	-	1.0	1.6	0.1	7.8

ENERGY SUPPLY / DEMAND BALANCE - 1995   **DENMARK**

million toe

	SOLID FUEL	PETROL PRODCT	NAT. GAS	NUCLR ENERGY	HEAT	HYDRO+ OTHERS	RENEW ENER	TOTAL
PRIM PRODUCT	–	4.0	2.5	–	–	–	0.2	6.7
TOT IMPORTS	9.0	7.7	–	–	–	0.1	–	16.8
TOT EXPORTS	–	1.4	0.7	–	–	–	–	2.1
STOCK CHANGE	–	–	–	–	–	–	–	–
GROSS CONSUM	9.0	10.3	1.8	–	–	0.1	0.2	21.4
BUNKERS	–	0.4	–	–	–	–	–	0.4
INLAND CONSM	9.0	9.9	1.8	–	–	0.1	0.2	21.0
ELEC POWER S	7.6	0.8	–	–	-1.5	-2.8	0.1	4.2
OTHER TRANSF	–	–	–	–	–	–	–	–
ENERG INDUST	–	0.8	0.0	–	0.2	0.4	–	1.4
FINAL CONSUM	1.4	8.3	1.8	–	1.3	2.5	0.1	15.4
– NON ENERGY	–	0.5	–	–	–	–	–	0.5
– INDUSTRY	0.7	1.5	0.6	–	0.2	0.7	–	3.7
– TRANSPORT	–	3.5	–	–	–	0.0	–	3.5
– HOUSEHOLD	0.7	2.8	1.2	–	1.1	1.8	0.1	7.7

ENERGY SUPPLY / DEMAND BALANCE - 2000

million toe

	SOLID FUEL	PETROL PRODCT	NAT. GAS	NUCLR ENERGY	HEAT	HYDRO+ OTHERS	RENEW ENER	TOTAL
PRIM PRODUCT	–	4.0	2.6	–	–	–	0.3	6.9
TOT IMPORTS	10.2	7.2	–	–	–	0.1	–	17.5
TOT EXPORTS	–	1.4	0.7	–	–	–	–	2.1
STOCK CHANGE	–	–	–	–	–	–	–	–
GROSS CONSUM	10.2	9.8	1.9	–	–	0.1	0.3	22.3
BUNKERS	–	0.4	–	–	–	–	–	0.4
INLAND CONSM	10.2	9.4	1.9	–	–	0.1	0.3	21.9
ELEC POWER S	8.6	0.8	–	–	-1.8	-3.1	0.1	4.6
OTHER TRANSF	–	–	–	–	–	–	–	–
ENERG INDUST	–	0.8	0.0	–	0.3	0.4	–	1.5
FINAL CONSUM	1.6	7.8	1.9	–	1.5	2.8	0.2	15.8
– NON ENERGY	–	0.5	–	–	–	–	–	0.5
– INDUSTRY	0.8	1.6	0.6	–	0.2	0.8	–	4.0
– TRANSPORT	–	3.6	–	–	–	0.1	–	3.7
– HOUSEHOLD	0.8	2.1	1.3	–	1.3	1.9	0.2	7.6

Summarized Energy Balance - DENMARK

in million toe	1973[a]	1980[a]	1983[a]	1990[b]	1995[b]	2000[b]
I. Gross Energy Consumption	20.21	19.53	16.63	20.70	21.40	22.30
- Bunkers	0.68	0.42	0.43	0.40	0.40	0.40
- Inland consumption	19.53	19.11	16.20	20.30	21.00	21.90
II. Inland Energy Consumption	19.53	19.11	16.20	20.30	21.00	21.90
- Solid fuels	2.32	5.84	5.46	8.40	9.00	10.20
- Oil	17.23	13.23	10.37	10.00	9.90	9.40
- Gas	-	-	-	1.70	1.80	1.90
- Primary electricity. etc.	-0.02	0.04	0.37	0.20	0.30	0.40
III.Indigenous Production[1]	0.07	0.30	2.18	6.50	6.70	6.90
- Hard coal	-	-	-	-	-	-
- Lignite & peat	-	-	-	-	-	-
- Oil	0.07	0.30	2.18	4.00	4.00	4.00
- Natural gas	-	-	-	2.40	2.50	2.60
- Nuclear energy	-	-	-	-	-	-
- Hydro & geothermal[2]	0.00	0.00	0.00	-	-	-
- Others & renewables	-	-	-	0.10	0.20	0.30
IV. Net Imports[3]	20.28	19.25	14.23	14.20	14.70	15.40
- Solid fuels	1.96	6.01	5.37	8.40	9.00	10.20
- Oil	18.34	13.20	8.50	6.40	6.30	5.80
- Natural gas[2]	-	-	-	-0.70	-0.70	-0.70
- Electricity[2]	-0.02	0.04	0.36	0.10	0.10	0.10
V. Stock changes[4]	0.14	0.02	-0.22	-	-	-
- Solid fuels	-0.36	0.17	-0.09	-	-	-
- Oil	0.50	-0.15	-0.13	-	-	-
- Gas	-	-	-	-	-	-
VI. Electricity Generation Input	4.66	6.24	5.40	7.90	8.50	9.50
- Solid fuels[5]	1.82	5.06	5.15	7.20	7.60	8.60
- Oil	2.84	1.18	0.25	0.70	0.80	0.80
- Natural gas	-	-	-	-	-	-
- Nuclear energy	-	-	-	-	-	-
- Hydro & geothermal[2]	0.00	0.00	0.00	-	-	-
- Others & renewables	-	-	-	0.00	0.10	0.10

Main indicators (related to long term objectives)

	1973 - 1963	1980 -1975	1983 -1980	1990 -1983	2000 -1990
Inland energy annual growth rates	+5.3%	+1.8%	-5.4%	+3.3%	+0.8%
GDP annual growth rates	+4.5%	+2.7%	+1.4%	+2.0%	+2.6%
Energy-GDP ratio	1.18	0.67	-3.86	1.65	0.31

	1973	1980	1983	1990	1995	2000
Share of oil in gross energy consumption	88.6%	69.9%	64.9%	50.2%	48.1%	43.9%
Share of coal and nuclear in in electricity production	39.1%	81.1%	95.4%	91.1%	89.4%	90.5%
Supply dependance on imports	100%	98.6%	85.6%	68.6%	68.7%	69.1%

Sources: a. Statistical Office of the European Communities
         b. Energy 2000
Notes: 1. Production of primary sources, including recovered products.
         2. The conversion of electricity, including hydro and geothermal. is
            based on its actual energy content: 3600 kjoules/kWh or 860 kcal/kWh
         3. The (-) sign means net exports
         4. The (-) sign means a stock decrease
         5. Including coke oven gas and blast furnace gas (derived from coal)

G E R M A N Y

A.   ECONOMIC REFERENCE FRAMEWORK

1.   For the period 1983-2000, the GDP growth assumption is an average
rate of 2.7% per annum. Growth would be slightly higher in the 1990s
at 2.8% p.a. against 2.6% during the period 1983-90.

From 1983 to 1990, public consumption and public investment will both
grow at the same low rate of 0.7% per annum. But the increase in
wages in real terms (1.5% per annum) will boost private consumption,
which will increase at an average annual rate of 1.9%. GDP growth
will also be fuelled by a 3.5% p.a. growth in gross fixed capital
formation. The marked increase in profits and the recovery in world
trade accounts for the growth in business investment. Finally, the
highly competitive position of manufacturing industry in a more
favourable world environment will help to push up exports (3.6% per
annum).

The trends assumed for the period up to 1990 continue on average over
the period 1991-2000. The GDP growth rate will be close to 3% over
this period. A continuing feature of the German economy will be its
healthy export capacity, particularly in the capital goods
industries.

At the sectoral level, the assumptions adopted give the following
results:

- consumer goods industries such as agri-foodstuffs and textiles
  benefit from the buoyancy of private consumption;

- the building industry and non-metal mineral products show moderate
  success, the net result of two opposing factors, namely, the low
  level of public investment and the upswing in housebuilding;

- the capital goods industries show fairly strong growth throughout
  the period, encouraged by the high level of business investment and
  good export prospects;

- the chemical industry also benefits from the growth in world trade,
  while steel output continues close to 1981 levels throughout the
  period;

- the service sector's share of GDP continues to increase, rising
  from 54% in 1983 to 56% in 2000.

Overall, growth will be led mainly by the service sector and the
capital goods industries, which explains the increasing shares of
these sectors in the value added generated by the productive system,
as shown in Table 1.

Table 1: Sectoral trends

	1970	AAGR[13]	1980	AAGR	1983	AAGR	1990	AAGR	2000
							DM '000 million at 1975 prices		
Agriculture	27.2	1.5	31.6	4.7	36.3	-0.2	35.8	0.0	35.8
Industrie	389.9	2.0	476.8	-1.0	462.2	2.6	554.6	2.7	723.2
EIIs[14]	101.5	2.4	129.0	-1.9	121.6	2.2	141.4	1.2	159.5
Other industries	288.4	1.9	347.8	-0.7	340.6	2.8	413.2	3.2	563.7
Building	63.9	2.1	78.8	-2.9	72.1	2.2	83.8	2.5	107.3
Services	466.5	3.6	661.8	0.9	679.4	2.7	816.7	3.0	1097.6
Total value added (GDP)	947.5	2.8	1249.0	0.0	1250.0	2.6	1490.9	2.8	1963.9

B.   TOTAL FINAL ENERGY DEMAND

2.   Total final energy demand will increase up to the year 2000 at a low average annual rate (0.4%) compared with GDP (2.7%). This contrast in the overall trends needs to be qualified where individual sectors are concerned: energy demand for household consumption (residential sector and passenger transport) will decrease over the period to 2000, but this will be offset by a significant increase in the demand for energy in the production sector (industry, tertiary sector and freight transport).   Two general phenomena explain these trends: the fall,   albeit slight,   in the population and industry's dynamism, putting its stamp on economic activity.

Nevertheless,   real energy savings will be made.   Overall energy intensity[15] will decrease by 32% over a period of 20 years, from 161 (000 toe/DM 1 000 million (1975) in 1980 to 129 in 1990 and 99 in 2000.

The effects of structural change in the economic fabric and the housing stock, and technical improvements in methods of manufacture, processing plant and machinery and construction will all contribute to this energy saving.

3.   One of the most significant effects of these changes is the increase in specific end-uses of electricity in industry and the tertiary sector: the capital goods and consumer goods industries will make intensive use of electricity as they achieve higher industry ranking; and the productivity gains obtained in the tertiary sector will mean an increase in electricity consumption (development of data transmission and processing systems, increasing sophistication and proliferation of electrical equipment).

For other fuels,   the gradual shift away from oil products is confirmed,   while gas consumption increases slightly and the level of coal consumption is maintained.   District heating continues to make slow progress without exceeding 3% of total demand.

---

[13] Average annual growth rate.
[14] EEIs: Energy-intensive industries = metals, non-metal minerals, paper, chemicals.
[15] Intensity is defined here as the ratio of total final energy demand to GDP.

(a) Industry

4.   Despite fairly vigorous activity in manufacturing industry (average annual growth rate of 2.7% up to 2000), energy demand in the sector increases only slowly (0.8%/year); this increase is attributable to specific electricity and, to a lesser extent, natural gas.

Structural changes are the first reason for this trend: the shift in industrial activity towards light industry (capital goods and consumer goods industries) will boost steam end-uses, the share of which in total heating end-uses will increase from 40% at present to 47% in 2000; it will also entail a rapid increase in specific electricity, in particular through the expansion of electro-mechanics. The more sluggish activity of the energy-intensive industries will lead to a corresponding reduction in furnace end-uses.

A second reason is the introduction of new processes, more rational use of energy and more efficient conversion processes in the supply of heat and steam: for example, energy consumed per tonne of steel produced will decrease by 20%, and consumption per tonne of cement and glass will fall by 30%. For the steam and furnace end-uses of light industries, the improvement would be about 20% and 25% respectively.

Overall, the energy intensity of industry will decline from 136 (000 toe/DM 000 million (1975) in 1980 to 106 in 1990 and 85 in 2000, a fall of 37% in 20 years.

5.   Structural changes in energy end-uses will make for significant changes in the final energy balance of the industrial sector by 2000. In particular, the accelerating advance of specific electricity will bring it up to second place in 2000, with 27% of the market, behind gas (32%), consumption of which will be only slightly above 1983 levels.

Coal consumption will be boosted by an expansion of the steam market and, to a lesser extent, by the development of combined steam/power production units. Electricity will continue to make headway in furnace end-uses.

The consumption of oil products in 2000 will be down to 1983 levels, 46% less than in 1980.

(b) Transport

6.   Energy consumption in the transport sector will increase very slowly (0.2% per year). The marked increase in goods transport, thanks to the economic buoyancy of manufacturing industry, will merely make up for declining consumption in passenger transport.

Consumption in passenger transport will decline over the period to 2000, reversing the trend of the last 20 years. The switch to public transport is the principal source of energy savings despite the increase in total passenger traffic: the share of public transport would rise from 31 to 37% for urban traffic and from 20 to 24% for inter-city traffic. Improvements in the specific consumption of cars and better organization of town traffic also explain this trend: they will reduce average vehicle consumption by approximately 20%.

7.   The steady growth rate for goods transport (1.8% on average from 1980 to 2000) is mainly attributable to road haulage, with its 70% share in total energy consumption by goods transport. Overall, the share of goods transport in the total consumption of the sector increases significantly, from 39% in 1980 to 48% in 2000.

(c) Residential and tertiary sector

8.   Energy consumption in the residential sector, which is very dependent on the number of housing units and the type of construction, will be favourably affected by two factors which will reduce it slightly between now and 2000: the number of households will increase only slightly, and thermally efficient new dwellings will quickly supersede older housing.

Energy consumption in the tertiary sector will increase as the sector experiences vigorous expansion.   The increase will not be in consumption for heating purposes,   which will decline,   but in consumption of specific electricity.

Overall, the energy consumption of the residential and tertiary sector (including agriculture) will amount to 69 million toe in 2000, i.e. 4 million toe less than in 1983. As a result of this relative stability, the energy intensity[16] of the sector in 2000 will be 42% below the 1980 figure: 35 (000 toe/DM 000 million (1975) in 2000, as against 47 in 1990 and 61 in 1980.

9.   The total number of housing units will increase by only 3% over 20 years, without any change in the breakdown by type: one-family houses will continue to account for 47% of the total up to 2000. On the other hand, the average age of the buildings rapidly declines, approximately half of the total in 2000 being made up of dwellings less than 25 years old.

Central heating then reaches its widest spread, which will have an unfavourable impact on the average energy consumption per dwelling. But it will facilitate the introduction of stricter standards for insulation (leading to 15% heating savings per new dwelling in 2000) and an improvement in the efficiency of equipment, two elements which will reduce unit consumption. These latter effects will carry the day, for the average final energy demand for heating per dwelling will be 20% lower in 2000 than today.

Despite the economic upswing in the tertiary sector, the number of office jobs and thus the total volume of offices will remain very stable between now and 2000. As office buildings will be replaced at much the same rate as housing, their insulation will be improved,

16 Energy intensity is defined here as the ratio of the energy demand of the residential and tertiary sector (expressed in million toe) to GDP (expressed in DM 10^7 at 1975 prices).

leading to a reduction in useful-heat requirements. The same applies to educational establishments. For the distributive trades whose buildings are not energy-efficient, heating requirements will be slightly greater.

Specific electricity will have a large share in the electricity demand of the sector, accounting for three quarters of the total in 2000, as a result of an annual 3% increase in requirements. Most of this increase will be in the tertiary sector.

10. The energy balance for the residential and tertiary sector will not show any marked changes. The share of heating oil will gradually decline, to be supplanted by electricity (mainly specific electricity) and, to a lesser extent, district heating and natural gas. The advance of electricity is mainly due to specific electricity, but also to a slight increase in thermal uses for new housing.

Table 2:   Final energy consumption

	Million toe				Annual percentage changes	
	1980	1983	1990	2000	1990/83	2000/90
Industry	65.0	53.7	58.6	61.3	+1.3	+0.5
Transport	40.4	40.5	41.6	42.2	+0.4	+0.1
Residential & tertiary	75.8	73.3	71.2	69.4	−0.4	−0.3
Non-energy uses	19.4	17.8	20.0	21.0	+1.7	+0.5
Total	200.7	182.1	191.6	193.9	+0.7	+0.1
− solid fuels	20.4	17.7	16.8	16.7	−0.7	−0.1
− oil products	114.8	99.5	102.5	95.2	+0.4	−0.7
− gas	35.6	34.6	35.7	38.3	+0.4	+0.7
− electricity	27.5	27.9	32.4	37.8	+2.2	+1.6
− other[17]	2.4	2.4	4.2	5.9	+8.3	+3.5

C. ENERGY SUPPLY

11. Table 3, showing gross inland primary energy consumption (GIC) based on the economic scenario adopted, indicates an increase in GIC of about 1.5% per annum until 1990 and 0.5% per annum thereafter, making an average annual rate of growth of 0.8% over the period 1983-2000.

The major structural feature is the increased share of nuclear power, which rises to 12.4% in 1990 and 17.5% in 2000 against 6.7% in 1983. This change is mainly at the expense of oil products, whose share is reduced to 39.8% in 1990 and 35.4% in 2000, against 43.3% in 1983.

The nuclear programme and the exploitation of indigenous solid fuel and oil and gas resources enables Germany to reduce its dependence on imported energy below 50% despite the increase in primary energy consumption. A closer analysis of these aggregate energy supply trends reveals movements which are much more marked.

---

[17] Heat and renewables.

Table 3: Gross inland primary energy consumption

(million toe)

Form of energy	1983	1990	1995	2000
Coal products	80.1	83.8	87.6	88.4
Oil products	106.5	109.0	105.6	101.4
Natural gas	39.6	43.3	41.8	42.1
Nuclear fuels	16.5	34.1	42.8	50.3
Primary electricity	2.4	2.1	2.1	2.1
Renewable energy sources	0.9	1.4	1.8	2.4
GIC	245.9	273.8	281.7	286.7

(a) Oil products

12. Despite their declining share in GIC, consumption of oil products remains relatively stable throughout the period. This apparent stability is the net result of several trends:

- lower sales of heavy fuel oil to the electricity sector, in which its use is kept to a minimum;

- significantly lower sales of heating oil to the residential and tertiary sector (28.7 million toe in 1990 and 22.8 million toe in 2000 compared with 36.2 million toe in 1982), heating oil losing market shares to natural gas, district heating networks and electricity;

- increased diesel fuel deliveries to the transport sector because of the growing number of diesel motor vehicles and the increase in demand by goods transport;

- steady sales of light products: increased deliveries of petrochemical cuts and kerosene compensate for lower sales of petrol.

13. These trends together with stable total consumption of oil products mean that the share of light products in the consumption structure (Table 4) will hold steady.

Table 4: Structure of consumption of oil products

%

Product	1980	1983	1990	1995	2000
Gaseous products	1.82	1.85	2.52	2.46	2.19
Light products	28.66	34.25	34.47	34.54	33.55
Medium products	44.90	45.69	43.65	42.67	43.12
Heavy products	24.65	18.21	19.36	20.24	21.15

Considering that the Federal Republic of Germany will remain a net importer of oil products within the limits observed between 1975 and 1983, the changes in the consumption structure are bound to be reflected almost directly in the production structure. Consequently, there is a clear need to continue the development of conversion units (catalytic cracking, hydrocracking and coking) in order to make sure that refining capacity is adapted to requirements.

14. On the basis of the refining capacity in service on 1 January 1983,
     i.e. 123 million tonnes of atmospheric distillation, 22.4 million
     tonnes of cracking and coking plus 5.7 million tonnes of visbreaking,
     Germany is generally well equipped, given some investment in
     conversion units, to cope with the medium-term structural changes and
     to ensure that its oil industry as a whole has a utilization rate of
     over 75%, thus ensuring its profitability. However, simple
     topping-reforming   refineries still account for 35 million tonnes
     of distillation capacity, nearly 28% of the total capacity of the
     country. Except where particular local conditions offer major outlets
     for heavy fuel oil, this type of refinery is no longer suited to the
     structure of the market. These refineries must therefore be the first
     target for investment in conversion units or else they may have to be
     closed, which would make Germany all the more dependent on outside
     supplies for the purchase of its refined products.

(b) Natural gas

15. The share of natural gas in GIC remains stable (approximately 15%)
     throughout the period. That implies an appreciable increase by 1990
     in the volumes of natural gas imported (31.0 million toe in 1990,
     compared with 27.8 million toe in 1982). However, that corresponds
     only to the minimum purchases to be made under the contracts signed
     to date. After 1990, if the exploitation of national resources
     continues at present levels, imports would remain stable, which could
     necessitate a new contract for about 4 to 5 million toe by 2000.
     Since the intention is to restrict the share supplied by any one
     supplier to 30% of inland consumption, imports from Norway and/or the
     Netherlands would be needed to cover this additional demand.

16. The pattern of inland consumption of natural gas will continue to
     reflect present trends, namely more buoyant sales to the residential
     and tertiary sectors (despite competition from district heating
     networks and more locally from electricity for thermal uses),
     consolidation of the share acquired in the industrial markets and a
     slight reduction in the long run of sales to power stations.

     It should be stressed that the introduction of strict pollutant
     emission standards for all large combustion plants could consolidate
     the penetration of natural gas on the industrial markets and in the
     electricity sector.

(c) Coal products

17. Consumption of coal products, a plentiful national resource, will
     rise moderately until the end of the period, their share in GIC
     falling only slightly from 32.6% in 1983 to 30.3% in 1990 and 30.5%
     in 2000, on the assumption that the intention is to consume an
     abundant national resource in order to minimize the country's
     dependence on imports. It is in fact for this reason that the
     Government has for a long time encouraged the use of coal in
     conventional power stations as well as for the combined production of
     heat and electricity both in the industrial sectors and for the heat
     distribution networks of district heating schemes.

18. The growth of electricity consumption, though moderate, accounts for the bulk of the increase in overall consumption (55.6 million toe in 1990 and 59.4 million toe in 2000 compared with 54.9 million toe in 1983), while consumption by the steel industry remains fairly stable. Sales to other industries increase only slightly, since the major conversions to coal (as in cement factories) have already been completed.

19. National lignite resources will continue to be exploited until the end of the period, almost the entire output going to the electricity sector.

20. Lastly, despite all current research efforts, mainly in the United States but also in the Federal Republic of Germany, it seems likely that, given the expected pattern of energy prices, coal-conversion techniques (gasification and liquefaction) will not become economic until after 2000.

(d) Electricity sector

21. The main feature of the electricity sector is the increasing share of nuclear power in meeting demand, which will grow at an annual rate of 2.2% between 1983 and 1990 and 1.6% between 1991 and 2000, making an average annual growth rate of 1.8% over the period.

22. Medium- and long-term capital investment policy is based on a link between coal and nuclear energy. For the latter, the expiry in 1982 of the moratorium on new building permits should enable a 28 GW capacity to be installed by 1995. If new construction is impossible for political reasons, the resulting additional cost in terms of the average price of the kWh would certainly affect the penetration of electricity on some of its markets, including heating end-uses in the residential and tertiary sector.

Table 5: Capital investment programme for the electricity sector

Units	1981–1990	1991–1995	1996–2000
Nuclear (PWR)	13040 MWe	5580 MWe	4910 MWe
Nuclear (HTR)	296 MWe	–	–
Nuclear (FBR)	292 MWe	–	–
Coal-fired plants	7050 MWe	4765 MWe	2525 MWe
Three-fuel plants	726 MWe	2860 MWe	3751 MWe
Hydro	784 MWe	–	–
Renewables[18]	278 MWe	189 MWe	321 MWe

---

[18] Wind, photovoltaic, incinerators.

23. The overall production structure (all producers) by type of fuel
    shows the large share of nuclear power: from 1990 onwards, over 30%
    of the electricity generated is of nuclear origin.   Although its
    relative share falls, electricity produced from solid fuels will
    increase considerably in volume terms (by 10.5% from 1983 to 2000).
    In the short term, the share of liquid fuels in electricity
    generation becomes marginal, while gaseous fuels, though on a
    downward trend after 1990, keep a significant share.

Table 6: Total net electricity production

	1983		1990		1995		2000	
Origin	TWh	%	TWh	%	TWh	%	TWh	%
Hydro	18.6	5.3	18.8	4.5	17.0	3.7	17.0	3.4
Nuclear	62.4	17.8	135.4	32.4	169.6	37.4	199.5	41.3
Solid fuels	211.1	60.3	218.4	52.2	233.2	51.4	233.5	48.5
Liquid fuels	13.3	3.8	3.3	0.8	3.0	0.7	2.5	0.5
Gas. fuels	41.1	11.6	41.2	9.8	29.7	6.5	27.6	5.7
Other	4.1	1.2	1.2	0.3	1.6	0.3	2.6	0.6
Total	351.5	100.0	418.2	100.0	454.1	100.0	482.7	100.0

TWh and %

D. SUMMARY TABLES

ENERGY SUPPLY/DEMAND BALANCE   1980   **GERMANY**

million toe

	SOLID FUEL	PETROL PRODCT	NATURL GAS	NUCLR ENERGY	HEAT	HYDRO+ OTHERS	RENEW ENER	TOTAL
PRIM PRODUCT	88.7	5.0	14.3	11.1	0	1.5	1.0	121.4
TOT IMPORTS	8.1	138.4	32.6	0.0	0	1.7	0.0	180.8
TOT EXPORTS	13.5	7.2	2.0	0.0	0	1.2	0.0	23.8
STOCK CHANGE	-0.7	-4.5	-0.2	0.0	0	0.0	0.0	-5.4
GROSS CONSUM	82.7	131.7	44.6	11.1	0	2.0	1.0	273.0
BUNKERS	0.0	2.9	0.0	0.0	0	0.0	0.0	2.9
INLAND CONSM	82.7	128.9	44.6	11.1	0	2.0	1.0	270.1
ELEC POWER S	49.9	5.6	16.1	11.1	-2.4	-30.1	1.0	51.1
OTHER TRANSF	10.3	1.9	-10.5	0.0	0	0.0	0.0	1.6
ENERG INDUST	0.6	8.1	3.5	0.0	0	4.6	0.0	16.7
FINAL CONSUM	20.4	114.8	35.6	0.0	2.4	27.5	0.0	200.7
- NON ENERGY	1.2	16.8	1.3	0.0	0	0.0	0.0	19.4
- INDUSTRY	13.4	18.3	19.8	0.0	0.8	12.7	0.0	65.0
- TRANSPORT	0.1	39.4	0.0	0.0	0	0.9	0.0	40.4
- HOUSEHOLD	5.7	40.2	14.5	0.0	1.6	13.9	0.0	75.8
STAT DIFFER	1.5	-1.5	0.0	0.0	0	0.0	0.0	0.0

ENERGY SUPPLY/DEMAND BALANCE   1983   **GERMANY**

million toe

	SOLID FUEL	PETROL PRODCT	NATURL GAS	NUCLR ENERGY	HEAT	HYDRO+ OTHERS	RENEW ENER	TOTAL
PRIM PRODUCT	83.7	4.3	13.6	16.5	–	1.5	0.9	120.5
TOT IMPORTS	8.6	110.0	27.8	–	–	2.0	–	148.4
TOT EXPORTS	10.1	7.3	1.5	–	–	1.1	–	20.0
STOCK CHANGE	-1.0	3.5	-0.3	–	–	–	–	2.2
GROSS CONSUM	81.2	110.5	39.6	16.5	–	2.4	0.9	251.1
BUNKERS	–	2.5	–	–	–	–	–	2.5
INLAND CONSM	81.2	108.0	39.6	16.5	–	2.4	0.9	248.6
ELEC POWER S	54.3	2.7	9.6	16.5	-2.4	-30.5	0.9	51.2
OTHER TRANSF	8.8	0.1	-7.9	–	–	–	–	1.0
ENERG INDUST	0.4	5.7	3.3	–	–	4.9	–	14.3
FINAL CONSUM	17.7	99.5	34.6	–	2.4	27.9	–	182.1
– NON ENERGY	0.0	15.9	1.9	–	–	–	–	17.8
– INDUSTRY	13.0	9.8	17.6	–	0.8	12.5	–	53.7
– TRANSPORT	0.0	39.7	–	–	–	0.8	–	40.5
– HOUSEHOLD	4.2	33.4	15.5	–	1.6	14.6	–	73.3
STAT. DIFFER	0.4	0.7	-0.3	–	–	–	–	0.8

ENERGY SUPPLY/DEMAND BALANCE – 1990

million toe

	SOLID FUEL	PETROL PRODCT	NATURL GAS	NUCLR ENERGY	HEAT	HYDRO+ OTHERS	RENEW ENER	TOTAL
PRIM PRODUCT	82.3	5.0	13.8	34.1	–	1.2	1.4	137.8
TOT IMPORTS	9.5	113.0	31.0	–	–	2.2	–	155.7
TOT EXPORTS	8.0	6.1	1.5	–	–	1.3	–	16.9
STOCK CHANGE	–	–	–	–	–	–	–	–
GROSS CONSUM	83.8	111.9	43.3	34.1	–	2.1	1.4	276.7
BUNKERS	–	2.9	–	–	–	–	–	2.9
INLAND CONSM	83.8	109.0	43.3	34.1	–	2.1	1.4	273.8
ELEC POWER S	55.6	0.4	10.3	34.1	-3.9	-34.0	1.1	63.6
OTHER TRANSF	10.5	0.6	-7.0	–	–	–	–	4.1
ENERG INDUST	0.9	5.5	4.3	–	–	3.7	–	14.4
FINAL CONSUM	16.8	102.5	35.7	–	3.9	32.4	0.3	191.6
– NON ENERGY	1.0	17.7	1.3	–	–	–	–	20.0
– INDUSTRY	12.2	13.1	17.7	–	1.4	14.3	–	58.6
– TRANSPORT	–	40.8	–	–	–	0.8	–	41.6
– HOUSEHOLD	3.6	30.9	16.7	–	2.5	17.3	0.3	71.2

ENERGY SUPPLY/DEMAND BALANCE   1995   **GERMANY**

million toe

	SOLID FUEL	PETROL PRODCT	NATURL GAS	NUCLR ENERGY	HEAT	HYDRO+ OTHERS	RENEW ENER	TOTAL
PRIM PRODUCT	79.1	5.0	13.8	42.8	–	1.2	1.8	143.7
TOT IMPORTS	16.5	109.5	29.5	–	–	2.2	–	157.7
TOT EXPORTS	8.0	6.0	1.5	–	–	1.3	–	16.8
STOCK CHANGE	–	–	–	–	–	–	–	–
GROSS CONSUM	87.6	108.5	41.8	42.8	–	2.1	1.8	284.6
BUNKERS	–	2.9	–	–	–	–	–	2.9
INLAND CONSM	87.6	105.6	41.8	42.8	–	2.1	1.8	281.7
ELEC POWER S	59.2	0.4	6.7	42.8	-4.6	-37.1	1.4	68.8
OTHER TRANSF	10.3	0.5	-6.8	–	–	–	–	4.0
ENERG INDUST	0.8	5.7	4.4	–	–	3.8	–	14.7
FINAL CONSUM	17.3	99.0	37.5	–	4.6	35.4	0.4	194.2
– NON ENERGY	1.1	18.0	1.4	–	–	–	–	20.5
– INDUSTRY	13.3	11.8	18.8	–	1.6	15.3	–	60.7
– TRANSPORT	–	42.1	–	–	–	0.9	–	43.0
– HOUSEHOLD	2.9	27.1	17.3	–	3.0	19.2	0.4	69.9

ENERGY SUPPLY/DEMAND BALANCE   2000

million toe

	SOLID FUEL	PETROL PRODCT	NATURL GAS	NUCLR ENERGY	HEAT	HYDRO+ OTHERS	RENEW ENER	TOTAL
PRIM PRODUCT	77.7	4.0	13.8	50.3	–	1.2	2.4	139.4
TOT IMPORTS	18.7	106.1	29.8	–	–	2.2	–	156.8
TOT EXPORTS	8.0	5.8	1.5	–	–	1.3	–	16.6
STOCK CHANGE	–	–	–	–	–	–	–	–
GROSS CONSUM	88.4	104.3	42.1	50.3	–	2.1	2.4	289.6
BUNKERS	–	2.9	–	–	–	–	–	2.9
INLAND CONSM	88.4	101.4	42.1	50.3	–	2.1	2.4	286.7
ELEC POWER S	59.4	0.4	6.0	50.3	-5.1	-39.6	1.6	73.0
OTHER TRANSF	11.5	0.4	-6.6	–	–	–	–	5.3
ENERG INDUST	0.8	5.4	4.4	–	–	3.9	–	14.5
FINAL CONSUM	16.7	95.2	38.3	–	5.1	37.8	0.8	193.9
– NON ENERGY	1.2	18.3	1.5	–	–	–	–	21.0
– INDUSTRY	13.2	10.9	19.3	–	1.6	16.3	–	61.3
– TRANSPORT	–	41.2	–	–	–	1.0	–	42.2
– HOUSEHOLD	2.3	24.8	17.5	–	3.5	20.5	0.8	69.4

Summarized Energy Balance - GERMANY

in million toe	1973[a]	1980[a]	1983[a]	1990[b]	1995[b]	2000[b]
I. Gross Energy Consumption	265.79	272.99	251.05	276.70	284.60	289.60
- Bunkers	3.58	2.85	2.46	2.90	2.90	2.90
- Inland consumption	262.21	270.14	248.59	273.80	281.70	286.70
II. Inland Energy Consumption	262.21	270.14	248.59	273.80	281.70	286.70
- Solid fuels	83.16	82.67	81.20	83.80	87.60	88.40
- Oil	146.21	128.86	108.00	109.00	105.60	101.40
- Gas	26.99	44.59	39.62	43.30	41.80	42.10
- Primary electricity, etc.	5.85	14.02	19.77	37.60	46.70	54.80
III. Indigenous Production[1]	119.18	121.43	120.52	137.80	143.70	149.40
- Hard coal	69.13	62.19	58.44	58.60	53.90	52.50
- Lignite & Peat	22.92	26.50	25.24	25.20	25.20	25.20
- Oil	7.14	4.97	4.31	5.00	5.00	4.00
- Natural gas	15.02	14.25	13.65	13.80	13.80	13.80
- Nuclear energy	3.05	11.06	16.46	34.10	42.80	50.30
- Hydro & geothermal[2]	1.20	1.49	1.46	1.20	1.20	1.20
- Others & renewables	0.72	0.97	0.96	1.40	1.80	2.40
IV. Net Imports[3]	147.41	156.96	128.37	138.80	140.90	140.20
- Solid fuels	-10.15	-5.37	-1.51	1.50	8.50	10.70
- Oil	144.65	131.25	102.65	106.90	103.50	100.30
- Natural gas[2]	12.03	30.58	26.33	29.50	28.00	28.30
- Electricity[2]	0.88	0.50	0.89	0.90	0.90	0.90
V. Stock changes[4]	0.81	5.40	-2.16	-	-	-
- Solid fuels	-1.27	0.65	0.97	-	-	-
- Oil	2.01	4.51	-3.50	-	-	-
- Gas	0.07	0.24	0.36	-	-	-
VI. Electricity Generation Input	69.50	85.14	85.47	102.70	112.70	118.90
- Solid fuels[5]	47.02	52.62	55.89	58.20	61.70	61.90
- Oil	9.75	5.64	2.71	0.40	0.40	0.40
- Natural gas	7.76	13.36	7.99	7.70	5.20	3.50
- Nuclear energy	3.05	11.06	16.46	34.10	42.80	50.30
- Hydro & geothermal[2]	1.20	1.49	1.46	1.20	1.20	1.20
- Others & renewables	0.72	0.97	0.96	1.10	1.40	1.60

Main indicators (related to long term objectives)

	1973 -1963	1980 -1975	1983 -1980	1990 -1983	2000 -1990
Inland Energy annual growth rates	+4.3%	+2.4%	-2.7%	+1.4%	+0.5%
GDP annual growth rates	+4.5%	+3.5%	0.0%	+2.6%	+2.8%
Energy-GDP ratio	0.96	0.69	-	+0.54	+0.18

	1973	1980	1983	1990	1995	2000
Share of oil in gross energy consumption	56.4%	48.2%	44.0%	40.4%	38.1%	36.0%
Share of coal and nuclear in electricity production	72.0%	74.8%	84.6%	89.9%	92.7%	94.4%
supply dependance on imports	55.5%	57.5%	51.5%	50.2%	49.5%	48.4%

Sources: a. Statistical Office of the European Communities
          b. Energy 2000

Notes: 1. Production of primary sources, including recovered products.
      2. The conversion of electricity, including hydro and geothermal, is based on its
         actual energy content: 3600 kjoules/kWh or 860 kcal/kWh
      3. The (-) sign means net exports
      4. The (-) sign means a stock decrease
      5. Including coke oven gas and blast furnace gas (derived from coal)

G R E E C E

A.   ECONOMIC REFERENCE FRAMEWORK

1.   For the period 1983-2000 the GDP growth assumption is an average
annual rate of 2.6%. This growth will be spread unevenly over that
period: 2.4% from 1983 to 1990, and 2.8% from 1991 to 2000.

Between 1983 and 1990, general growth will be stimulated mainly by
public consumption (+3.0% per annum on average) and public
investment (+7.1% per annum on average), to the detriment of the
public deficit.   Private consumption will increase by 1.6% per
annum, in line with the real increase in wages (+1.6% per annum).
Although gross fixed capital formation will benefit only slightly
from the investment in housebuilding (+1.2% per annum), it will
nevertheless increase at a rate of 3.2% as a result of public
investment and investment in capital goods.   Despite a fairly
favourable international climate, exports will not rise by more
than 2.5% per annum.

The trends observed over the period 1983 to 1990 will continue in
much the same vein over the period 1991 to 2000, with perhaps a
slight upswing and a 2.7% growth in GDP over that period.

At the sectoral level, certain major trends stand out:

- rapid growth in the building trade due to the increase in public
  investment;

- average growth in the services sector;

- steady growth in manufacturing industry.

These trends are illustrated in Table 1.

Table 1: Sectoral trends

Unit: 1000 million DRA 1975								
	1970	AAGR	1980	AAGR	1983	AAGR	1990	AAGR	2000
Agriculture	71.2	4.3	108.9	0.7	111.2	1.8	126.4	2.0	154.1
Manufacturing	94.2	4.9	151.5	-1.5	145.0	3.4	183.6	2.9	245.5
Building	31.3	4.3	47.9	0.1	48.0	5.4	69.2	4.0	102.4
Services	214.1	5.0	348.2	0.1	349.5	1.6	390.0	2.7	506.9
Total value added (GDP)	410.8	4.8	656.5	-0.1	653.7	2.4	769.2	2.7	1008.9

B. TOTAL FINAL ENERGY DEMAND

2. Greece is a special case in the reference projection since there is as yet no MEDEE demand-analysis model for that country. But although the methodology used is less precise than for the other countries, the information was sufficient to establish a projection of the final energy and non-energy demand. It must be said, however, that the final-energy consumption levels were not established on the basis of a detailed analysis of useful-energy requirements and that the figures are therefore not as reliable.

3. The total final energy demand will increase by 2.9% per annum between 1983 and 2000. By comparison, over the same period the GDP growth rate will be 2.6% per annum. These broadly similar rates of in energy consumption and GDP are explained by the increase in the pace of industrialization in Greece and the rise in the standard of living. As a result of these factors, consumption in the various sectors (industry, residential and tertiary, and transport) will grow at roughly the same rate, although the increase will be greater in industry in the period until 1990.

   Between 1983 and 2000, the breakdown of final energy consumption will be marked mainly by a slight reduction in the share of oil products, dropping from 75.4% in 1983 to 70.1% in 2000, to the benefit of solid fuels, whose share will increase from 7.5% to 8.7%, and electricity, which, making considerable headway in the residential and tertiary sectors, will account for 17.9% of the total in 2000 as against 15.8% in 1980. After 1990, gas will also make further inroads both in the residential sector and in industry.

4. Unlike in the other Community countries, manufacturing industry in Greece will see its share in the total value added increase from 27.9% in 1980 to 32.1% in 2000, reflecting the increase in the pace of industrialization. Over the period from 1983 to 2000, energy consumption in industry will increase by an average of 3.6% per annum. This rise will mainly concern oil products (+1.3 million toe), imported solid fuels (+0.8 million toe) and electricity (+0.7 million toe). Although increasing in volume, oil products will account for less than half (48%) of total consumption by the year 2000, compared with 53% in 1983 and 65% in 1980.

5. Passenger and goods transport will develop at about the same rate, although for different reasons. As the carriage of goods is closely bound up with economic activity, industrial development should generate considerable growth in this form of transport. Passenger transport will continue to be mainly car-orientated, and the number of vehicles will increase rapidly with the rise in the standard of living.

   Energy consumption in the transport sector should therefore increase (+2.1 million toe in 17 years), and diesel oil's share should rise significantly.

6. Per-capita energy consumption in the residential and tertiary sectors in Greece is the lowest in the entire Community, and as a result of the favourable climate only a small proportion is for the heating of buildings. In the residential sector the increase in the standard of living will make for greater energy consumption, in particular as a result of greater use of to air-conditioning systems. The expansion of the services sector (average of +2.2% per annum from 1983 to 2000) will also boost energy consumption. Overall consumption in the residential and tertiary sectors will increase by 2.5% per annum over the period from 1983 to 2000, mainly as a result of increased consumption of electricity whose share will rise from 33.3% to 39.1%.

Table 2:   Final energy consumption

	Million toe				Annual percentage changes	
	1980	1983	1990	2000	1990/83	2000/90
Industry	4.0	3.6	5.2	6.6	5.4	2.4
Transport	3.9	4.3	5.2	6.4	2.8	2.1
Residential and tertiary	2.7	3.0	3.4	4.6	1.8	3.1
Non-energy uses	0.6	0.5	0.6	0.8	2.6	2.9
Total	11.2	11.4	14.4	18.4	3.4	2.5
− solid fuels	0.6	0.8	1.2	1.6	6.0	2.9
− oil products	8.9	8.6	10.6	12.9	3.0	2.0
− gas	0.0	0.1	0.1	0.6	0.0	19.6
− electricity	1.7	1.8	2.5	3.3	4.8	2.8

C. ENERGY SUPPLY

7. Table 3, which shows the trend in gross inland primary energy consumption (GIC) arising from this scenario, indicates that GIC will increase by 2.9% between 1983 and 2000. Over the same period, GDP will increase by an average of 2.6% per annum, giving a GIC/GDP elasticity of 1.11, compared with 1.33 over the period from 1975 to 1983.

The main feature in terms of structure is the increase in the share of solid fuels from 22% in 1980 to 29% in 1983 and 40% as from 1990.   Oil products are the principal victim of this change, their share dropping to 55% in the year 2000 as against 76% in 1980 and 69% in 1985.

The use of indigenous resources for the production of electricity (lignite, hydroelectricity and new technologies) will help to reduce the dependence on imports from 84.1% in 1980 to 63.1% in the year 2000.

A source-by-source analysis of these aggregate energy supply trends reveals movements that are much more marked.

Table 3:   Gross inland primary energy consumption (million toe)

Form of energy	1983	1990	1995	2000
Coal products	4.5	8.0	8.6	10.0
Oil products	10.7	11.5	12.8	13.8
Natural gas	0.1	0.1	0.2	0.6
Hydro and geothermal	0.4	0.4	0.6	0.8
Renewables	0.0	0.1	0.1	0.1
GIC	15.6	20.0	22.3	25.3

a) Oil products

8. The increase in the consumption of oil products (+2.1 mtoe in 17 years despite a reduction in their proportionate share of GIC) is not, of course, uniform across the entire range of products. On the contrary, the following features emerge:

    - significant increase in the sale of heavy fuel oil to industry, despite the inroads made by solid fuels;

    - sharp drop in the sale of residual fuel oil to power stations, which after 1990 should have to use oil or gas only for standby generation;

    - shift in the transport sector towards diesel oil in an expanding market;

    - upsurge in petrochemicals and thus in naphtha supplies;

    - slight increase in the use of heating oil in the tertiary and residential sectors.

    Overall, the breakdown of consumption and production will reflect a trend towards light and medium products during the period under consideration.

Table 4:   Structure of consumption of oil products (%)

Products	1980	1983	1990	1995	2000
Gaseous products	3.3	2.0	3.0	2.8	2.8
Light products	22.0	28.1	24.8	26.8	27.3
Medium products	29.2	33.1	34.5	33.1	33.2
Heavy products	45.5	36.9	37.7	37.2	36.7
	100.0	100.0	100.0	100.0	100.0

b)   Natural gas

9. In the long run, natural gas will make only a small contribution to covering the energy needs of the country (2% in the year 2000). As indigenous production of natural gas will remain low, it will be necessary to import from Algeria, Italy or the USSR. A direct link

with Algeria would undoubtedly be uneconomic because of the low level of consumption expected and the cost of installing the link. The cost of liquefying natural gas and transporting it by methane tankers would also be prohibitive given the expected demand.

c)   Coal products

10. The increase in the share of solid fuels in GIC (+6.6 million toe from 1980 to 2000) is due to the vast programme to exploit the national lignite resources in connection with a progressive change in the electricity-generating sector towards this fuel and to massive inroads by imported steam coal in the industrial sector.

Between 1983 and the year 2000, deliveries to power stations will increase by 4.8 million toe and deliveries to the industrial sector by 0.8 million toe. Lignite production will rise from 3.8 million toe in 1983 to 7.3 million toe in 1990 and 9 million toe in the year 2000.

d)   Electricity-generating sector

11. The main feature of the electricity-generating sector is obviously lignite's increasing share in meeting demand which will rise at an annual rate of 4.6% between 1983 and 1990 and 2.8% between 1991 and the year 2000, i.e. an annual average growth rate of 3.6% over the period considered.

12. Investments will be channelled mainly into lignite-fired power stations (+2500 MW up to 1990, +1300 MW between 1991 and 2000) and into hydroelectric and geothermal power stations (+2300 MW over the period considered).

Nine 300 MW lignite-fired power stations are currently planned for the period 1983-1992, as are nine hydroelectric units with a total capacity of 1730 MW. The reference scenario has not taken into account the construction of nuclear power stations.

If the Greek authorities were to opt for nuclear power this would not have an effect until after 1995. A 900 or 1200 MW power station would undoubtedly be involved. The forecasts for the installation of lignite-fired power stations would have to be revised downwards in that case.

13. The breakdown of total production by type of fuel underlines the high proportionate share of lignite since, after 1990, more than three quarters of all the electricity produced will be generated from this fuel.

Liquid fuels will be the hardest hit by competition from lignite, their use being gradually confined to standby generation.

Table 5 : Total net electricity production (in TWh and in %)

Origin	1983		1990		1995		2000	
	TWh	%	TWh	%	TWh	%	TWh	%
Hydro and geothermal	2.3	10.5	4.3	13.5	6.6	18.4	8.4	19.8
Solid fuels	13.7	61.4	26.3	82.5	28.1	78.5	32.8	77.4
Liquid fuels	6.3	28.1	1.1	3.4	0.9	2.4	0.9	2.1
Others	0.0	0.0	0.2	0.6	0.2	0.7	0.3	0.7
Total	22.3	100	31.9	100	35.8	100	42.4	100

D. SUMMARY TABLES

ENERGY SUPPLY / DEMAND BALANCE - 1980   **GREECE**

(Mio Toe)	SOLID FUEL	PETROL PRODCT	GAS	NUCLR ENERGY	HEAT	HYDRO+ OTHERS	RENEW ENER	TOTAL
PRIM PRODUCT	3.1	–	–	–	–	0.3	0.0	3.4
TOT IMPORTS	0.4	23.6	–	–	–	0.0	–	24.0
TOT EXPORTS	–	10.6	–	–	–	0.0	–	10.6
STOCK CHANGE	-0.1	-0.6	–	–	–	–	–	-0.7
GROSS CONSUM	3.4	12.4	–	–	–	0.3	0.0	16.1
BUNKERS	–	0.8	–	–	–	–	–	0.8
INLAND CONSM	3.4	11.6	–	–	–	0.3	0.0	15.3
ELEC POWER S	2.5	2.1	–	–	–	-1.7	0.0	2.9
OTHER TRANSF	0.1	0.1	-0.1	–	–	–	–	0.1
ENERG INDUST	–	0.5	0.0	–	–	0.3	–	0.8
FINAL CONSUM	0.6	8.9	0.0	–	–	1.7	–	11.2
– NON ENERGY	–	0.6	–	–	–	–	–	0.6
– INDUSTRY	0.5	2.6	0.0	–	–	0.9	–	4.0
– TRANSPORT	–	3.9	–	–	–	0.0	–	3.9
– HOUSEHOLD	0.1	1.8	0.0	–	–	0.8	–	2.7
STAT DIFFER	0.2	0.0	–	–	–	–	–	0.2

ENERGY SUPPLY / DEMAND BALANCE — 1983   **GREECE**

(Mio Toe)	SOLID FUEL	PETROL PRODCT	GAS	NUCLR ENERGY	HEAT	HYDRO+ OTHERS	RENEW ENER	TOTAL
PRIM PRODUCT	3.6	1.2	0.1	−	−	0.2	0.0	5.2
TOT IMPORTS	0.9	16.2	−	−	−	0.2	−	17.3
TOT EXPORTS	−	6.3	−	−	−	0.0	−	6.4
STOCK CHANGE	−	0.4	−	−	−	−	−	0.3
GROSS CONSUM	4.5	11.5	0.1	−	−	0.4	0.0	16.4
BUNKERS	−	0.8	−	−	−	−	−	0.8
INLAND CONSM	4.5	10.7	0.1	−	−	0.4	0.0	15.6
ELEC POWER S	3.6	1.5	−	−	−	−1.8	0.0	3.2
OTHER TRANSF	0.1	0.2	−	−	−	−	−	0.2
ENERG INDUST	−	0.4	0.0	−	−	0.4	0.0	0.8
FINAL CONSUM	0.8	8.6	0.1	−	−	−1.8	−	11.4
− NON ENERGY	−	0.4	0.1	−	−	−	−	0.5
− INDUSTRY	0.8	1.9	−	−	−	0.8	−	3.6
− TRANSPORT	−	4.3	−	−	−	−	−	4.3
− HOUSEHOLD	0.0	2.0	−	−	−	1.0	−	3.0
STAT DIFFER	−	−	−	−	−	−	−	−

ENERGY SUPPLY / DEMAND BALANCE — 1990

(Mio Toe)	SOLID FUEL	PETROL PRODCT	GAS	NUCLR ENERGY	HEAT	HYDRO+ OTHERS	RENEW ENER	TOTAL
PRIM PRODUCT	7.3	0.3	0.1	−	−	0.4	0.1	8.2
TOT IMPORTS	0.7	22.2	−	−	−	−	−	22.9
TOT EXPORTS	−	9.5	−	−	−	−	−	9.5
STOCK CHANGE	−	−	−	−	−	−	−	−
GROSS CONSUM	8.0	13.0	0.1	−	−	0.4	0.1	21.6
BUNKERS	−	1.5	−	−	−	−	−	1.5
INLAND CONSM	8.0	11.5	0.1	−	−	0.4	0.1	20.1
ELEC POWER S	6.8	0.3	−	−	−	−2.5	0.1	4.7
OTHER TRANSF	−	0.2	−	−	−	−	−	0.2
ENERG INDUST	0.0	0.4	−	−	−	0.4	−	0.8
FINAL CONSUM	1.2	10.6	0.1	−	−	2.5	−	14.4
− NON ENERGY	−	0.6	−	−	−	−	−	0.6
− INDUSTRY	1.2	2.8	−	−	−	1.2	−	5.2
− TRANSPORT	−	5.2	−	−	−	0.0	−	5.2
− HOUSEHOLD	−	2.0	0.1	−	−	1.3	−	3.4

ENERGY SUPPLY / DEMAND BALANCE - 1995   **GREECE**

(Mio Toe)	SOLID FUEL	PETROL PRODCT	GAS	NUCLR ENERGY	HEAT	HYDRO+ OTHERS	RENEW ENER	TOTAL
PRIM PRODUCT	7.8	0.3	0.1	−	−	0.6	0.1	8.9
TOT IMPORTS	0.8	23.5	0.1	−	−	−	−	24.4
TOT EXPORTS	−	9.5	−	−	−	−	−	9.5
STOCK CHANGE	−	−	−	−	−	−	−	−
GROSS CONSUM	8.6	14.3	0.2	−	−	0.6	0.1	23.8
BUNKERS	−	1.5	−	−	−	−	−	1.5
INLAND CONSM	8.6	12.8	0.2	−	−	0.6	0.1	22.3
ELEC POWER S	7.2	0.2	−	−	−	-2.7	0.1	4.8
OTHER TRANSF	−	0.2	−	−	−	−	−	0.2
ENERG INDUST	0.0	0.4	−	−	−	0.5	−	0.9
FINAL CONSUM	1.4	12.0	0.2	−	−	2.8	−	16.4
− NON ENERGY	−	0.7	−	−	−	−	−	0.7
− INDUSTRY	1.4	3.1	0.1	−	−	1.3	−	5.9
− TRANSPORT	−	5.9	−	−	−	0.0	−	5.9
− HOUSEHOLD	−	2.3	0.1	−	−	1.5	−	3.9

ENERGY / SUPPLY DEMAND BALANCE - 2000

(Mio Toe)	SOLID FUEL	PETROL PRODCT	GAS	NUCLR ENERGY	HEAT	HYDRO+ OTHERS	RENEW ENER	TOTAL
PRIM PRODUCT	9.0	0.0	0.0	−	−	0.8	0.1	9.9
TOT IMPORTS	1.0	24.8	0.6	−	−	−	−	26.4
TOT EXPORTS	−	9.5	−	−	−	−	−	9.5
STOCK CHANGE	−	−	−	−	−	−	−	−
GROSS CONSUM	10.0	15.3	0.6	−	−	0.8	0.1	26.8
BUNKERS	−	1.5	−	−	−	−	−	1.5
INLAND CONSM	10.0	13.8	0.6	−	−	0.8	0.1	25.3
ELEC POWER S	8.4	0.2	−	−	−	-3.1	0.1	5.6
OTHER TRANSF	−	0.2	−	−	−	−	−	0.2
ENERG INDUST	0.0	0.5	−	−	−	0.6	−	1.1
FINAL CONSUM	1.6	12.9	0.6	−	−	3.3	−	18.4
− NON ENERGY	−	0.8	−	−	−	−	−	0.8
− INDUSTRY	1.6	3.2	0.3	−	−	1.5	−	6.6
− TRANSPORT	−	6.4	−	−	−	0.0	−	6.4
− HOUSEHOLD	−	2.5	0.3	−	−	1.8	−	4.6

Summarized Energy Balance – GREECE

in million toe	1973[a]	1980[a]	1983[a]	1990[b]	1995[b]	2000[b]
I. Gross Energy Consumption	12.51	16.12	16.40	21.60	23.80	26.80
– Bunkers	0.83	0.83	0.80	1.50	1.50	1.50
– Inland consumption	11.68	15.29	15.60	20.10	22.30	25.30
II. Inland Energy Consumption	11.68	15.29	15.60	20.10	22.30	25.30
– Solid fuels	2.19	3.35	4.46	8.00	8.60	10.00
– Oil	9.28	11.57	10.67	11.50	12.80	13.80
– Gas	–	–	0.07	0.10	0.20	0.60
– Primary electricity, etc.	0.21	0.37	0.40	0.50	0.70	0.90
III.Indigenous Production[1]	1.97	3.35	5.19	8.20	8.90	9.90
– Hard coal	–	–	–	–	–	–
– Lignite & peat	1.77	3.03	3.64	7.30	7.80	9.00
– Oil	–	–	1.24	0.30	0.30	–
– Natural gas	–	–	0.07	0.10	0.10	–
– Nuclear energy	–	–	–	–	–	–
– Hydro & geothermal[2]	0.19	0.29	0.20	0.40	0.60	0.80
– Others & renewables	0.01	0.03	0.04	0.10	0.10	0.10
IV. Net Imports[3]	11.60	13.55	10.90	13.40	14.90	16.90
– Solid fuels	0.46	0.39	0.88	0.70	0.80	1.00
– Oil	11.13	13.11	9.86	12.70	14.00	15.30
– Natural gas[2]	–	–	–	–	0.10	0.60
– Electricity[2]	0.01	0.05	0.16	–	–	–
V. Stock changes[4]	1.07	0.78	-0.31	–	–	–
– Solid fuels	0.04	0.07	0.06	–	–	–
– Oil	1.03	0.71	-0.37	–	–	–
– Gas	–	–	–	–	–	–
VI. Electricity Generation Input	3.41	4.94	5.32	7.60	8.10	9.50
– Solid fuels[5]	1.47	2.52	3.57	6.80	7.20	8.40
– Oil	1.74	2.10	1.51	0.30	0.20	0.20
– Natural gas	–	–	–	–	–	–
– Nuclear energy	–	–	–	–	–	–
– Hydro & geothermal[2]	0.19	0.29	0.20	0.40	0.60	0.80
– Others & renewables	0.01	0.03	0.04	0.10	0.10	0.10

Main indicators (related to long term objectives)

	1973–1963	1980–1975	1983–1980	1990–1983	2000–1990
Inland Energy annual growth rates	+13.0%	+5.4%	+0.7%	+3.7%	+2.3%
GDP annual growth rates	+7.7%	+4.4%	+0.0%	+2.4%	+2.7%
Energy-GDP ratio	1.69	1.22	–	1.54	0.85

	1973	1980	1983	1990	1995	2000
Share of oil in gross energy consumption	80.8%	76.9%	69.9%	60.2%	60.1%	57.1%
Share of coal and nuclear in in electricity production	43.1%	51.0%	67.1%	89.5%	88.9%	88.4%
Supply dependance on imports	92.7%	84.1%	66.5%	62.0%	62.6%	63.1%

Sources: a. Statistical Office of the European Communities
         b. Energy 2000
Notes: 1. Production of primary sources, including recovered products.
         2. The conversion of electricity, including hydro and geothermal, is
            based on its actual energy content: 3600 kjoules/kWh or 860 kcal/kWh
         3. The (-) sign means net exports
         4. The (-) sign means a stock decrease
         5. Including coke oven gas and blast furnace gas (derived from coal)

F R A N C E

A. ECONOMIC REFERENCE FRAMEWORK

1. For the period 1983-90 the working assumption is that growth in the French economy will be moderate because at the beginning of the period France must restore its foreign trade balance. However, a recovery in profit margins will subsequently enable undertakings to consider new investments and, combined with a fairly sustained economic recovery in European countries and an upswing in household consumption, this should boost national production in the long run. Between 1983 and 1990 the average annual GDP growth rate should be 2.2%.

The trends observed in the latter years of the period 1983-90 will be more or less carried over into the period 1991-2000, and the average annual GDP growth rate should increase to 2.8%.

At the sectoral level, the assumptions made produce the following results:

- An atmosphere more conducive to investment will benefit the capital goods industry (mechanical engineering and, particularly, electrical equipment and electronics), but the revival of these industries will be offset to some extent by imports as the investments that will be made will be in sectors using a relatively large proportion of imports. As a result, the growth rate in these industries will be between 2.5 and 3% a year (i.e. significantly higher than in other industries).

- Sustained demand abroad after 1985 will revive French exports, in particular of non-ferrous metals, the products of the chemical and allied industries, agri-foods and certain capital goods. An increase in exports of this type will also lead to increased activity in the transport sector and engineering services. These trends reflect sectoral changes which will operate to the advantage of these industries.

- An upturn in household consumption will further benefit consumer durables, the motor industry, and particularly services, rather than the more traditional food and clothing sectors.

Generally speaking, services and capital goods will provide the main impetus for growth, which explains why these industries will account for a growing proportion of value added as shown in Table 1.

Table 1: Sectoral trends

Unit: 1000 million FF									
	1970	AAGR[18]	1980	AAGR	1983	AAGR	1990	AAGR	2000
Agriculture	50.6	1.2	57.2	1.6	60.0	1.0	64.3	1.0	71.0
Industry	245.8	3.8	355.7	0.4	359.6	1.8	408.8	2.1	501.3
EEI[19]	64.5	3.0	86.6	1.1	89.5	1.2	97.6	1.6	114.9
Other industries	181.3	4.0	269.1	0.1	270.1	2.0	311.2	2.2	386.4
Building	58.2	-0.1	57.9	-1.1	56.0	0.5	58.1	1.0	64.2
Services	354.7	4.2	536.1	1.3	557.2	2.7	671.8	3.6	956.5
Total value added (GDP)	709.3	3.6	1006.9	0.9	1032.8	2.2	1203.0	2.8	1593.0

B.   TOTAL FINAL ENERGY DEMAND

2. Growth in total final energy demand between 1983 and 2000 will be relatively low (0.7% per annum on average) compared with a 2.6% per annum growth rate for economic activity.

This general uncoupling of energy demand from economic growth is encountered in all the major economic sectors, although the reasons differ between the different sectors.

- In the industrial sector, the structural changes to the benefit of light industry will reduce the proportion of energy consumption accounted for by the energy-intensive industries and will encourage more efficient thermal uses.

- In the transport sector, the switch towards using public transport, and the manufacture of vehicles with a lower specific consumption will reduce the fuel requirements of private cars.

- In the residential and tertiary sector economies in heating thanks to better insulation, more effective energy management and greater energy efficiency will compensate for the increase in the total number of housing units.

A marked expansion in electricity will fundamentally alter the current market shares of the various energy sources. Electricity's share in covering final energy demand will rise from 15.2% in 1983 to 23.0% in the year 2000, the biggest inroads being made in heating in the residential and tertiary sector. Gas consumption will also increase, but not as much as electricity consumption, rising by 17.2% over the period as a whole.

These switches will be at the expense of oil products, whose share in covering total thermal end-uses will fall by over 50% between 1983 and the year 2000.

----

[18] AAGR = average annual growth rate.

[19] EEI: energy-intensive industries = metals, non-metallic minerals, paper, chemicals.

(a) Industry

3. Major structural changes in industry will mean that energy demand
in industry will not match the increases in that sector's economic
activity.   Whilst the rate of increase in the value added by
industry as a whole will be 2.0% per annum between 1983 and 2000,
the rate of increase in energy consumption is no more than 1.0% per
annum.

This difference will lower industry's energy intensity from
126 (10^3 toe/10^7 FF (1970)) in 1980 to 97 in 1990 and 81 in the
year 2000, i.e. by 35% in 20 years.

A feature of the structural effects is a decrease in the amount of
energy consumed by the intermediate goods industries which are the
ones most affected by the structural changes.   In particular, there
will be a big decrease (20% in 17 years) in the demand for fuels
for furnaces,   the biggest source of energy demand in these
industries. This retreat by the energy-intensive industries will be
to the advantage of the capital goods and consumer goods
industries,   which are characterized from the energy point of view
by a greater recourse to steam.   The steam requirements of these
industries in 2000 will be 30% up on 1985.

Added to these effects resulting from structural changes will be
better technical coefficients resulting from switching to more
energy efficient processes.   For instance, in 2000 greater use of
scrap will give fuel savings of 35% per tonne of steel produced
(although this will be balanced by higher electricity consumption).
The amount of steam used per unit of value added in light
industries will decrease by 12% between now and 2000, as a result
of the more efficient use of energy.

4. Overall, final energy consumption for thermal purposes will remain
unchanged between 1983 and 2000 but steam end-uses in this field
will come to play a dominant part as their share increases from 46%
in 1983 to 56% in 2000. Oil products will lose a significant share
of the heat market, decreasing from 42% to 24%. Gas will remain
stable and coal's share will increase by 4%.

Specific electricity consumption will increase by 1.5% a year.   In
2000 it will cover approximately 28% of total energy consumption by
industry as compared with 23% in 1983.

(b) Transport

5. The trend in passenger transport will differ quite considerably
from the trend in freight transport, the latter being more affected
by changes in economic activity.

Passenger transport will mostly be by private car, although in
urban areas there will be a distinct switch to public transport.
Whilst accounting for 73% of urban traffic in 1980, the private car
will account for only 63% in 2000.   There will be a similar switch
in inter-city transport where the share of traffic accounted for by
the private car will decrease from 82% in 1980 to 76% in 2000.

This switch to public transport (which will shorten the average annual distance covered by private cars), together with a decrease in the specific consumption of the vehicles, will result in a growth rate of only 0.3% per annum in total energy consumption in the passenger transport sector.

Yet this will hardly alter the proportion of energy consumption in passenger traffic as a whole accounted for by private cars - which will decrease by only 2% by 2000 (from 91 to 89%).

Energy consumption in freight transport will increase more rapidly (1.6% per annum) as it is more closely linked with economic activity. Most goods traffic will continue to be moved by road, with road haulage expanding their share of the market from 61 to 69% between 1980 and 2000. However, there is some uncertainty surrounding this trend as the factors involved are rather unpredictable.

Energy consumption in passenger transport as a proportion of total consumption in the transport sector will decrease slightly, from-6 62% in 1980 to 60% in 2000.

(c)   Residential and tertiary

6.   Growth in the economy as a whole and growth in the energy consumption of this sector will be markedly out of step. Whereas the rate for the first is 2.6% per annum on average for the period 1983-2000, the second, at around 0.5% for the same period, is very low.

The result is a major drop in the sector's energy intensity[20] from 50 (10^3 toe/10^9 FF (1970) in 1980 to 44 in 1990 and 34 in 2000, i.e. a gain of 30% in 20 years.

7.   The structure of the housing stock was balanced in 1980 as blocks of flats and houses each accounted for 50% of total housing, and this will still be the case in the year 2000, although houses will account for a bigger proportion of new dwellings in 2000 (56%). Approximately two thirds of all housing in 2000 will be new or will have been converted to central heating, complying with building standards such that a 20-30% saving in energy consumption is achieved by the end of the period, as compared with housing built in the period 1975-80. Nevertheless, although this trend theoretically favours energy saving, it is counteracted by two factors. One is that central heating will be almost universal by 2000 whereas it was installed in only 61% of housing units in 1980. The other is the fact that the new dwellings are generally more spacious (houses) and require more energy to heat than was the case with older housing, despite stricter construction standards.

Average useful-energy requirements for heating per housing unit will be very similar in 2000 to what they were in 1983. Nevertheless, improvements in the performance of heating equipment

---

[20] Energy intensity is defined here as the ratio of total energy consumption in the residential and tertiary sectors (in 10^3 toe) to GDP (in 10^9 FF at 1970 prices).

and better management of the equipment will keep demand for final energy used for heating down to 1983 levels, despite the increase in the number of dwellings.

Finally, energy for other purposes will account for a large proportion of the increase in the total energy consumption of the sector during the projection period, in particular electricity for specific uses, including air-conditioning, which will increase by 2.8% in the domestic sector, while energy consumption for domestic hot water will remain stable.

8. Given relatively sustained growth in the tertiary sector (3.2% per annum as against 2.6% for GDP between 1983 and 2000), the uncoupling of energy demand growth from GDP growth in the sector will be particularly marked as energy consumption will remain virtually constant between 1983 and 2000. This trend is mainly due to major savings in heating which will account for approximately 75% of the thermal needs of the tertiary sector. Heating requirements will even decrease by approximately 10% between 1983 and 2000, mainly because office buildings will be newer and more energy-efficient than the housing stock.

9. Electricity and gas will take the biggest shares of the heat market for new buildings. In multi-occupancy buildings their shares will be slightly reduced by district heating and, to a lesser extent, the heat pump. Finally, growth in demand for electricity for heating will average out at 6% per annum and that for gas at 1.5% per annum between now and 2000. By then each of these energy sources will cover 35% of total energy consumption in the residential and tertiary sectors.

The use of gas oil will decrease by approximately 5% per annum and its share of the market will be no more than 20% in 2000.

Table 2: Final energy consumption

F R A N C E	Million toe			Annual % changes	
	1983	1990	2000	1990/83	2000/90
Industry	34.4	39.6	40.8	+ 2.0	+ 0.4
Transport	32.7	35.8	38.8	+ 1.3	+ 0.8
Residential & tertiary	49.9	52.9	54.0	+ 0.8	+ 0.2
Non-energy uses	11.8	10.8	13.2	- 1.3	+ 2.0
Total	129.9	139.1	146.8	+ 1.0	+ 0.5
- solid fuels	8.9	11.4	13.7	+ 3.6	+ 1.9
- oil products	78.0	73.0	67.6	- 0.9	- 0.8
- gas	23.2	25.9	27.2	+ 1.6	+ 0.5
- electricity	19.8	26.0	33.8	+ 4.0	+ 2.7
- other[21]	..	2.8	4.5	..	+ 4.9

---

[21] Heat + renewable forms of energy.

C.   ENERGY SUPPLIES

10.   Table 3, which shows the trend in gross inland primary energy
consumption (GIC) arising from this scenario, indicates that GIC
will increase by about 2.4% per annum between 1983 and 1990 and
1.0% per annum thereafter, giving an average annual growth rate of
1.6% over the period 1983-2000.   It should, however, be noted that
this trend is affected by the increase in exports of electricity
from nuclear power stations.   If the effect of this is eliminated,
there would be a 2% per annum increase in GIC between now and 1990
and a 1.1% per annum increase thereafter.

The main structural feature is the increase in the share of nuclear
power to 35% in 1990 and 40% in 2000 as against 21% in 1983.   The
main loser by this change will be oil products, whose share will be
reduced from 39% in 1983 to 33% in 1990 and 35% by 2000. There will
be no major change in the market shares of other fuels.

Completion of the French nuclear-based electricity programme will
therefore reduce dependence on imports from 62% in 1983 to 54% in
1990 and 50% in 2000.   A source-by-source analysis of these
aggregate energy supply trends reveals movements that are much more
marked.

Table 3: Gross inland primary energy consumption (million toe)

Form of energy	1983	1990	1995	2000
Coal products	25.2	23.6	24.0	27.9
Oil products	87.1	81.2	79.9	77.1
Natural gas	22.4	26.5	26.8	26.9
Nuclear fuel	37.4	74.3	85.6	93.4
Hydro	4.9	2.7	3.8	4.4
Renewables[22]	0.2	1.3	1.8	2.5
GIC	177.2	209.6	221.9	232.2

(a)   Oil products

11.   Naturally, the gradual drop in consumption of oil products will not
be spread evenly over all products.   On the contrary, the likely
trends are as follows:

- a reduction in sales of heavy fuel oil to industry, because of
  competition in traditional markets from natural gas imported
  under contract by 1990, from coal, to which some further
  industrial users will return and electricity, which will be used
  to an increasing extent outside its specific markets;

- a reduction in the amount of gas oil used for heating in the
  residential and tertiary sectors as these sectors turn
  increasingly to natural gas, electricity and district heating

----

[22] Solar and geothermal energy, household waste, wind, biomass, wood.   In
the case of the latter, only the increase compared with 1983
(0.2 million toe in SOEC statistics) has been recorded so as to
maintain consistency with the situation described in 1983.

networks.   Nevertheless,   this significant reduction will be
partially offset by the gradual conversion of the transport
sector to diesel fuel;

- the stabilization of sales of light products where a higher
offtake of petrochemical cuts will compensate for lower sales of
motor fuel.

12.   The end-result of the overall decrease in the consumption of oil
products and the stabilization in sales of light products
(petrol + kerosene + naphtha) will be to increase the share of
light products in total consumption from 33.0% in 1983 to 33.4% in
1990 to 36.1% in 2000, whilst the share of medium products will
stabilize at 39% and that of heavy products will rise from 22.5%
in 1983 to 25.0% in 1990 and 21.7% in 2000.

If foreign trade in oil products continues along the same lines as
during the period 1975-83, it is clear that changes in the pattern
of consumption will have a fairly direct effect on production
(Table 4). Changes in that pattern therefore clearly underline the
need for the further development of conversion plant (catalytic
cracking, hydrocracking and coking units).

In addition, as less and less crude is processed in refineries, in
the short and medium term more refineries will have to be shut down
to ensure utilization rates capable of securing a profit for the
rest. Thus, on the basis of the atmospheric distillation capacity
in 1982, and without any later modification, the utilization rate
will be 56% in 1990. What is more, because of the streamlining of
the production structure, complex refineries equipped with
conversion units will be better able to adjust to demand. Except in
particular local conditions this will be the death knell for
"topping-reforming" refineries.

Table 4 - Structure of production of oil products (%)

	1980	1983	1990	1995	2000
Gaseous products	2.94	3.31	2.71	2.97	3.11
Light products	24.46	30.87	28.66	30.23	31.00
Medium products	38.68	36.90	39.38	39.65	39.51
Heavy products	33.92	28.92	29.25	27.15	26.38

(b)   Natural gas

13.   The results show that the volumes of natural gas available under
contract by 1990 will slightly exceed the country's own needs.
This will also apply in other Community countries, and is due to
the fact that the supply contracts were concluded during the
preceding decade when there was a fair amount of optimism about the
outlook as regards demand and consumer prices. Consequently, two
possibilities arise: renegotiating with the exporting countries the
clauses relating to contractual volumes or seeking the best
possible means of marketing the surplus on the home market.

Against this background, increasing quantities of natural gas
(13.5 million toe in 1990 and 14.3 million toe in 2000) will be
sold in that part of the residential and tertiary sectors
accessible to the natural gas transmission and distribution systems

in existence or being developed. The remainder of the gas available will be supplied to industry, which will thereby become the regulator of supply and demand in the case of natural gas.

(c)   Coal products

14.   In the medium term the consumption of coal products will decrease (23.6 million toe in 1990 as against 25.2 million in 1983) as a result of a drop in the amount of coal consumed by the two main users: the electricity sector will gradually switch over to nuclear power stations instead of using plants partly fired by coal, and the traditional iron and steel industry whose activity is on the decline. On the other hand, the development of heating networks and the gradual conversion of various industries to coal will mitigate these phenomena.

15.   By 2000, coal products will have recovered to the level of their 1983 share of gross inland primary energy consumption as a result of higher sales to the electricity sector. The big increase in demand for electricity, sustained by the spread of electric heating, will increase the share of the load curve accounted for by average-utilization-rate equipment, for which coal is the most economic fuel, and this will help coal to remain competitive. It should also be pointed out that on the combined heat and power market – both in industry and in heating networks – coal is gradually establishing itself as the economically ideal fuel despite the heavy investment associated with it.

(d)   Electricity

16.   The fundamental features of the electricity sector is the increasing share of nuclear power which will grow at an annual rate of 4.9% between 1983 and 1990 and 2.0% between 1991 and 2000, i.e. at an average annual growth rate of 3.2% for the period under review.

17.   Turning to investment, the nuclear programme due for completion by 1990 (resulting in a capacity of 49 960 MWe) is more than sufficient to meet the growing demand for power (Table 5). In fact, from a strictly economic viewpoint, the programme may even seem ambitious given the mere 4.3% annual growth in demand in the home market between 1983 and 1990. As a result the utilization rate of PWR plant will be no more than 63% in 1990, despite an assumed level of exports of 40 TWh. The industrial constraints weighing on the electricity investment programme will in fact still be felt to a considerable extent in the period 1991-95 when seven or eight new PWRs should be brought into service, bringing the nuclear plant utilization rate down to 58% in 1995. Exports are then expected to be around 32 TWh.

Table 5 - Capital investment programme for the electricity sector

	1981-1990		1991-1995		1996-2000	
PWRs	37750	MWe	11050	MWe	7510	MWe
FBRs	1200	MWe	-		-	
Hydro	4470	MWe	1920	MWe	50	MWe
Coal	2085	MWe	325	MWe	6700	MWe
CHP	940	MWth	1000	MWth	1000	MWth

18. The overall production structure (all producers), broken down by type of fuel, underlines the importance of nuclear power: from 1990 over 70% of all electricity production will be nuclear-based (Table 6). In the long run, it will be the liquid and gaseous fuels that bear the brunt of competition from nuclear energy as they are increasingly relegated to covering demand peaks. In the short term coal will also be phased out of base-load production as successive PWRs are brought into service, but after 1995 it will recover a share of the market equivalent to what it held in 1981, in plants with an average utilization rate.

Table 6: Total net electricity production (in TWh and in %)

Origin	1983		1990		1995		2000	
	Twh	%	Twh	%	Twh	%	Twh	%
Hydro	70.7	24.9	68.7	17.3	71.7	16.2	73.9	15.3
Nuclear	136.9	48.3	280.7	70.7	323.6	73.3	353.8	73.1
Solid fuels	52.0	18.3	31.1	7.8	31.8	7.2	45.5	9.4
Liquid fuels	15.0	5.3	6.4	1.6	5.2	1.2	3.7	0.8
Gaseous fuels	8.0	2.9	9.5	2.4	8.1	1.8	5.9	1.2
Other	0.9	0.3	0.8	0.2	1.1	0.3	1.3	0.2
Total	283.5	100.0	397.2	100.0	441.5	100.0	484.1	100.0

D. SUMMARY TABLES

ENERGY SUPPLY / DEMAND BALANCE - 1980   **FRANCE**

(Mio Toe)	SOLID FUEL	PETROL PRODCT	GAS	NUCLR ENERGY	HEAT	HYDRO+ OTHERS	RENEW ENER	TOTAL
PRIM PRODUCT	12.5	2.6	6.3	16.3	-	6.0	0.13	43.9
TOT IMPORTS	20.9	126.2	16.2	-	-	1.4	-	164.7
TOT EXPORTS	0.9	13.6	-	-	-	1.1	-	15.6
STOCK CHANGE	-1.5	-2.2	-0.9	-	-	-	-	-4.6
GROSS CONSUM	31.1	113.0	21.6	16.3	-	6.3	0.13	188.4
BUNKERS	-	3.9	-	-	-	-	-	3.9
INLAND CONSM	31.1	109.2	21.6	16.3	-	6.3	0.13	184.6
ELEC POWER S	14.2	10.6	2.9	16.3	-	-16.1	0.13	28.1
OTHER TRANSF	5.3	1.2	-4.9	-	-	-	-	1.6
ENERG INDUST	0.2	6.4	1.2	-	-	4.2	-	11.9
FINAL CONSUM	11.8	88.3	21.9	-	-	18.2	-	140.2
- NON ENERGY	0.3	9.6	2.2	-	-	-	-	12.1
- INDUSTRY	7.9	18.8	9.8	-	-	8.3	-	44.8
- TRANSPORT	-	31.1	-	-	-	0.6	-	31.7
- HOUSEHOLD	3.5	28.9	9.9	-	-	9.3	-	51.6
STAT DIFFER	0.4	2.6	0.6	-	-	-	-	2.8

ENERGY SUPPLY / DEMAND BALANCE - 1983   **FRANCE**

(Mio Toe)	SOLID FUEL	PETROL PRODCT	GAS	NUCLR ENERGY	HEAT	HYDRO+ OTHERS	RENEW ENER	TOTAL
PRIM PRODUCT	11.6	2.6	5.6	37.4	-	6.0	0.2	63.4
TOT IMPORTS	12.9	92.1	18.8	-	-	0.6	-	124.4
TOT EXPORTS	1.0	11.1	-	-	-	1.8	-	13.9
STOCK CHANGE	1.8	6.0	-2.0	-	-	-	-	5.8
GROSS CONSUM	25.2	89.6	22.4	37.4	-	4.9	0.2	179.7
BUNKERS	-	2.5	-	-	-	-	-	2.5
INLAND CONSM	25.2	87.1	22.4	37.4	-	4.9	0.2	177.2
ELEC POWER S	13.0	3.5	1.8	37.4	-	-19.4	0.2	36.5
OTHER TRANSF	3.2	1.2	-3.4	-	-	-	-	0.9
ENERG INDUST	0.1	4.4	0.8	-	-	4.5	-	9.9
FINAL CONSUM	8.9	78.0	23.2	-	-	19.8	-	129.9
- NON ENERGY	0.2	8.8	2.7	-	-	-	-	11.8
- INDUSTRY	6.0	11.6	9.0	-	-	7.8	-	34.4
- TRANSPORT	-	32.1	-	-	-	0.6	-	32.7
- HOUSEHOLD	2.8	24.7	11.1	-	-	11.4	-	49.9
STAT DIFFER.	-0.1	0.8	0.4	-	-	-	-	1.1

ENERGY / SUPPLY DEMAND BALANCE - 1990

(Mio Toe)	SOLID FUEL	PETROL PRODCT	GAS	NUCLR ENERGY	HEAT	HYDRO+ OTHERS	RENEW ENER	TOTAL
PRIM PRODUCT	9.0	4.5	2.9	74.3	-	5.7	1.3	97.7
TOT IMPORTS	14.6	86.4	23.6	-	-	0.5	-	125.1
TOT EXPORTS	-	6.7	-	-	-	3.5	-	10.2
STOCK CHANGE	-	-	-	-	-	-	-	-
GROSS CONSUM	23.6	84.2	26.5	74.3	-	2.7	1.3	212.6
BUNKERS	-	3.0	-	-	-	-	-	3.0
INLAND CONSM	23.6	81.2	26.5	74.3	-	2.7	1.3	209.6
ELEC POWER S	8.9	2.1	2.3	74.3	-1.7	-29.5	0.2	56.6
OTHER TRANSF	3.2	1.2	-2.5	-	-	-	-	1.9
ENERG INDUST	0.1	4.9	0.8	-	0.0	6.2	-	12.0
FINAL CONSUM	11.4	73.0	25.9	-	1.7	26.0	1.1	139.1
- NON ENERGY	0.3	8.5	2.0	-	-	-	-	10.8
- INDUSTRY	8.1	10.6	10.4	-	1.0	9.5	-	39.6
- TRANSPORT	-	35.0	-	-	-	0.8	-	35.8
- HOUSEHOLD	3.0	18.9	13.5	-	0.7	15.7	1.1	52.9

ENERGY SUPPLY / DEMAND BALANCE - 1995   **FRANCE**

(Mio Toe)	SOLID FUEL	PETROL PRODCT	GAS	NUCLR ENERGY	HEAT	HYDRO+ OTHERS	RENEW ENER	TOTAL
PRIM PRODUCT	8.8	4.5	3.0	85.6	–	6.0	1.8	109.7
TOT IMPORTS	15.2	85.1	23.8	–	–	0.5	–	124.6
TOT EXPORTS	–	6.7	–	–	–	-2.7	–	9.4
STOCK CHANGE	–	–	–	–	–	–	–	–
GROSS CONSUM	24.0	82.9	26.8	85.6	–	3.8	1.8	224.9
BUNKERS	–	3.0	–	–	–	–	–	3.0
INLAND CONSM	24.0	79.9	26.8	85.6	–	3.8	1.8	221.9
ELEC POWER S	8.0	2.2	1.5	85.6	-2.1	-33.4	0.2	62.0
OTHER TRANSF	2.9	1.6	-2.3	–	–	–	–	2.2
ENERG INDUST	0.1	5.7	0.8	–	0.1	7.0	–	13.7
FINAL CONSUM	13.0	70.4	26.8	–	2.0	30.2	1.6	144.0
– NON ENERGY	0.2	9.3	2.3	–	–	–	–	11.8
– INDUSTRY	8.8	9.9	10.0	–	1.0	10.4	–	40.1
– TRANSPORT	–	37.1	–	–	–	0.9	–	38.0
– HOUSEHOLD	4.0	14.1	14.5	–	1.0	18.9	1.6	54.1

ENERGY SUPPLY / DEMAND BALANCE - 2000

(Mio Toe)	SOLID FUEL	PETROL PRODCT	GAS	NUCLR ENERGY	HEAT	HYDRO+ OTHERS	RENEW ENER	TOTAL
PRIM PRODUCT	8.8	4.5	3.0	93.4	–	6.0	2.5	118.2
TOT IMPORTS	19.1	82.3	23.9	–	–	0.5	–	125.8
TOT EXPORTS	–	6.7	–	–	–	2.1	–	8.8
STOCK CHANGE	–	–	–	–	–	–	–	–
GROSS CONSUM	27.9	80.1	26.9	93.4	–	4.4	2.5	235.2
BUNKERS	–	3.0	–	–	–	–	–	3.0
INLAND CONSM	27.9	77.1	26.9	93.4	–	4.4	2.5	232.2
ELEC POWER S	11.1	1.3	1.0	93.4	-2.4	-36.8	0.3	67.9
OTHER TRANSF	2.8	1.6	-2.2	–	–	–	–	2.2
ENERG INDUST	0.3	6.6	0.9	–	0.1	7.4	–	15.3
FINAL CONSUM	13.7	67.6	27.2	–	2.3	33.8	2.2	146.8
– NON ENERGY	0.2	10.4	2.6	–	–	–	–	13.2
– INDUSTRY	9.1	8.8	10.3	–	1.0	11.6	–	40.8
– TRANSPORT	–	37.8	–	–	–	1.0	–	38.8
– HOUSEHOLD	4.4	10.6	14.3	–	1.3	21.2	2.2	54.0

Summarized Energy Balance – FRANCE

in million toe	1973[a]	1980[a]	1983[a]	1990[b]	1995[b]	2000[b]
I. Gross Energy Consumption	180.04	188.42	179.74	212.60	224.90	235.20
– Bunkers	5.32	3.87	2.54	3.00	3.00	3.00
– Inland consumption	174.72	184.55	177.20	209.60	221.90	232.20
II. Inland Energy Consumption	174.72	184.55	177.20	209.60	221.90	232.20
– Solid fuels	28.73	31.06	25.23	23.60	24.00	27.90
– Oil	123.85	109.16	87.11	81.20	79.90	77.10
– Gas	13.59	21.56	22.41	26.50	26.80	26.90
– Primary electricity, etc.	8.55	22.77	42.45	78.30	91.20	100.30
III. Indigenous Production[1]	34.31	43.88	63.39	97.70	109.70	118.20
– Hard coal	16.35	11.67	10.80	8.20	8.08	8.20
– Lignite & peat	0.85	0.84	0.82	0.80	0.72	0.60
– Oil	1.98	2.55	2.59	4.50	4.50	4.50
– Natural gas	6.32	6.32	5.58	2.90	3.00	3.00
– Nuclear energy	4.54	16.33	37.41	74.30	85.60	93.40
– Hydro & geothermal[2]	4.14	6.04	6.03	5.70	6.00	6.00
– Others & renewables	0.13	0.13	0.16	1.30	1.80	2.50
IV. Net Imports[3]	145.93	149.13	110.54	114.90	115.20	117.00
– Solid fuels	9.90	20.06	11.84	14.60	15.20	19.10
– Oil	128.73	112.65	81.02	79.70	78.40	75.60
– Natural gas[2]	7.56	16.15	18.83	23.60	23.80	23.90
– Electricity[4]	-0.26	0.27	-1.15	-3.00	-2.20	-1.60
V. Stock Changes[4]	0.20	4.59	-5.81	–	–	–
– Solid fuels	-1.63	1.51	-1.77	–	–	–
– Oil	1.54	2.17	-6.04	–	–	–
– Gas	0.29	0.91	2.00	–	–	–
VI. Electricity Generation	34.87	50.26	62.52	93.50	103.50	113.80
– Solid fuels[5]	8.60	15.81	14.46	9.60	8.70	12.40
– Oil	15.37	10.63	3.55	2.10	2.20	1.30
– Natural gas	2.09	1.32	0.91	1.60	0.80	0.40
– Nuclear energy	4.54	16.33	37.41	74.30	85.60	93.40
– Hydro & geothermal[2]	4.14	6.04	6.03	5.70	6.00	6.00
– Others & renewables	0.13	0.13	0.16	0.20	0.20	0.30

Main indicators (related to long term objectives)

	1973–1963	1980–1975	1983–1980	1990–1983	2000–1990
Inland energy annual growth rates	+5.6%	+3.1%	-1.4%	+2.4%	+1.0%
GDP annual growth rates	+5.5%	+3.3%	+0.8%	+2.1%	+2.8%
Energy-GDP ratio	1.02	0.94	-1.75	1.14	0.36

	1973	1980	1983	1990	1995	2000
Share of oil in gross energy consumption	71.7%	60.0%	49.9%	39.6%	36.9%	34.1%
Share of coal and nuclear in in electricity production	37.7%	63.9%	83.0%	89.7%	91.1%	93.0%
Supply dependance on imports	81.1%	79.1%	61.5%	54.0%	51.2%	49.7%

Sources: a. Statistical Office of the European Communities
b. Energy 2000 Study, DG XVII

Notes: 1. Production of primary sources, including recovered products.
2. The conversion of electricity, including hydro and geothermal, is based on its actual energy content: 3600 kjoules/kWh or 860 kcal/kWh
3. The (-) sign means net exports
4. The (-) sign means a stock decrease
5. Including coke oven gas and blast furnace gas (derived from coal)

I R E L A N D

A. ECONOMIC REFERENCE FRAMEWORK

1. For the period 1983 to 2000, the GDP growth assumption is an average annual rate of 3.5%; the rate will be 3.6% from 1983 to 1990 and 3.4% from 1990 to 2000.

From 1983 to 1990, internal and external components of final demand will affect general growth very differently; the former will grow only moderately while the latter will make an expanding contribution to net general growth. Because of the large public deficit, growth in consumer and public investment expenditure will be severely restricted, running at average annual rates of +0.8% and -1.2% respectively. Although gross fixed capital formation will be handicapped by inadequate public investment, it will nevertheless grow at 2.2% per annum, stimulated by investment in capital goods. The positive growth differential in foreign trade variables will increase during the period, exports rising by a average of 8.2% per annum compared with a rise of only 4% for imports.

The trends observed up to 1990 will broadly continue from 1991 to 2000, with annual GDP growth remaining firmly above the 3% mark.

At sectoral level, certain broad trends can be discerned:

- vigorous growth in manufacturing industry, which will reap most benefit from the excellent export performance;

- sustained growth in the service sector.

Table 1 gives details of these trends.

Table 1: Sectoral trends

	1970	AAGR[23]	1980	AAGR	1983	AAGR	IRL million at 1975 prices 1990	AAGR	2000
Agriculture	547.0	0.4	571.0	1.1	590.0	4.3	791.0	1.0	873.7
Industry	731.0	4.5	1136.2	2.2	1212.0	4.6	1663.1	4.8	2660.2
Building	214.0	4.5	332.8	0.2	335.0	2.8	405.5	2.5	519.1
Services	1102.0	5.2	1837.0	0.6	1869.4	2.9	2285.0	3.2	3131.0
Total value added (GDP)	2594.0	4.1	3877.0	1.1	4006.4	3.6	5144.6	3.4	7184.0

----

[23] AAGR = average annual growth rate.

B.   TOTAL FINAL ENERGY DEMAND

2.   Total final energy demand will grow faster than in most other
      Community countries (2.6% per annum between now and 2000), partly
      because of faster industrial growth and partly because of the
      continuing link between energy consumption in the residential
      sector and household incomes.

      Although in other countries the link between these last two
      indicators will be weakened, this is unlikely to happen in Ireland
      because the number of households will increase considerably and so
      will their income.   These factors will result in a significant
      increase   in   energy   consumption   in   the   residential   sector
      (1.2% per annum)   and   in   the   passenger   transport   sector
      (1.9% per annum)   sectors with   considerable room for energy
      consumption growth as saturation thresholds are far from being
      reached at the present time.

      In the manufacturing industries, energy consumption will rise even
      faster, increasing at an average rate of 4.2% per annum between
      1983 and 2000. This rate of increase will be due to vigorous
      industrial   activity   (a   growth   rate   of   approximately
      4.7% per annum).

3.   The proportions of total final demand covered by the various forms
      of energy will remain very stable, although there will be a slight
      shift from oil products to gas for heating applications in the
      residential and industrial sectors.

(a)   Industry

4.   The energy balance in the industrial sector will have changed
      appreciably by the year 2000, in both volume and structural terms.
      In terms of volume, total consumption will be approximately double
      the present level, although this increase will vary according to
      the form of energy.   In terms of structure, a slight drop in the
      share of consumption of oil products and solid fuels will be taken
      up by gas.

      The structure of energy use in terms of low-temperature and
      high-temperature applications will also change by 2000, as a
      result of changes in the industrial structure towards light
      industries, encouraging faster development of low-temperature
      applications (for steam), which will account for 85% of
      heat-raising applications in 2000 compared with 77% at present.
      The corresponding decline in energy use for high-temperature
      applications (furnaces) is due to the fact that the manufacture of
      products with a high energy content will merely remain stable, and
      specific consumption gains will be obtained thanks to more
      efficient production methods.

      The energy intensity of industry, defined as the ratio of its
      energy consumption to industrial value added, will fall from 1.4
      (1 000 toe/IRL million (1975)) in 1980 to 1.2 in 2000, a 14%
      improvement in 20 years.

(b)   Transport

5.   Passenger transport will expand rapidly (2.0% per annum on average), led primarily by private transport. An increase in vehicle ownership of 10 to 18%, depending on household income category, and a large increase in the number of households will mean that more journeys will be made, resulting in greater energy consumption. This will not be offset by major shifts to public transport, which will account for only an additional 4% of traffic in 2000.

6.   Goods transport will grow at a slower, although still appreciable, average annual rate (1.3%), the growth being due to the steady economic activity of light industries. Road haulage will account for most of this traffic.

Goods transport will account for 34% of total energy consumption of the transport sector in 2000 against 36% in 1980.

(c)   Residential and tertiary sector

7.   Energy consumption in the residential sector will reflect the large increase in the number of households and the rapid development of certain energy uses which will not be approaching saturation between now and 2000. Consumption by the tertiary sector will be sustained by its strong economic activity; and it is a sector with a large workforce occupying a lot of office space.

Taken together the growth rate in the residential and tertiary sectors will be 2.4% a year between now and the year 2000, one of the highest growth rates in the Community. However, major energy savings are to be expected since the energy intensity[24] of the sector will decrease by 34% over 20 years (0.70 in 1980, 0.58 in 1990 and 0.46 in 2000).

8.   The housing stock will grow to keep pace with the increase in the number of households, i.e. 30% more dwellings in 2000.

This rapid increase will encourage the building of new houses and flats and the installation of central heating. Thus 52% of the housing stock in 2000 will consist of dwellings less than 25 years old and 70% will be fitted with central heating (35% in 1980). One-family houses will continue to account for most of the housing stock.

The only real advantage of these developments as far as energy demand is concerned is that new houses should be of much better thermal design. Heat losses should be reduced by approximately 20%, reducing by the same percentage the unit heat requirements of houses, although this improvement will not be sufficient to offset the increase in the number of dwellings and useful-heat requirements will increase slightly in overall terms.

----

[24] Defined as the ratio of energy consumption in the residential and tertiary sector (in thousand toe) to GDP (in IRL million at 1975 prices).

The use of energy for other purposes will increase more rapidly. For example, the specific electricity demand and energy demand for domestic hot water will increase by approximately 3.7% a year between now and 2000.

The fast-growing economic activity of the tertiary sector will affect demand for energy in this sector just as the rapid population increase will affect energy consumption in the residential sector. Similar trends can be expected there too (except for domestic hot water).

9. As the stock consists chiefly of one-family houses, oil will remain in wide use as a heating fuel. It will even take a market share from solid fuels in the continuing substitution process between now and 2000.

Table 2: Final energy consumption

	1983	1990	2000	1990/83	2000/90
	million toe			Annual % change	
Industry	1.6	2.5	3.2	+6.6	+2.5
Transport	1.6	2.0	2.4	+3.2	+1.9
Residential and tertiary	2.2	3.0	3.3	+4.5	+1.0
Non-energy uses	0.4	0.6	0.6	+6.0	-
Total	5.9	8.1	9.5	+4.6	+1.6
Solid fuels	1.3	1.5	1.6	+2.1	+0.6
Oil products	3.4	4.7	5.4	+4.9	+1.4
Gas	0.4	1.1	1.4	+13.6	+2.4
Electricity	0.8	0.9	1.1	+1.7	+2.0
Other	-	-	-	-	-

C. ENERGY SUPPLY

10. Table 3 gives a breakdown of gross inland primary energy consumption (GIC) and shows how its components change over time.

Table 3: Gross inland primary energy consumption

Form of energy	1980	1983	1990	1995	2000
Coal products	2.0	2.0	3.3	3.7	4.3
Oil products	5.6	4.1	5.2	5.6	5.8
Natural Gas	0.7	1.8	1.4	1.6	1.7
Electricity	0.1	0.1	0.1	0.1	0.1
Renewable energy sources	-	-	0.1	0.1	0.2
GIC	8.4	7.9	10.1	11.1	12.1

Between 1983 and 2000 GIC increases at a rate of 2.5% per annum. It is important to note that this increase in consumption is accompanied by a diversification of sources of supply which will r-educe the share of oil products in GIC to 48% by the year 2000 compared with 67% in 1980 and 52% in 1983. By exploiting its reserves of natural gas and peat Ireland will be able to reduce its dependence on imported energy from 77% in 1980 to 72% in 2000.

A source by-source analysis of these aggregate energy supply trends reveals movements that are much more marked.

(a)   Oil products

11.   By 2000, the share of oil products, after a decline in the period 1980-83, will have reached a level slightly higher than that of 1980. This upward trend, stemming from a substantial increase in economic activity, is the net result of several factors:

- increased sales of heavy fuel oils to industry, covering part of-t the increase in their energy consumption;

- lower sales of residual fuel oil to power stations as they convert to solid fuels;

- higher demand in the transport sector (up by 0.7 million toe between 1980 and 2000;

- increased sales of heating oil to the residential and tertiary sectors as their total energy demand rises and in partial substitution for peat as a heating fuel.

Because the reduction in sales of heavy oil products to power stations will in the long term be offset by an increase in deliveries to industry, their share in the consumption structure will increase from 33.9% in 1983 to 38.1% in 2000. As the increases in the consumption of light and medium products will be similar, their respective shares will be unchanged in the year 2000. It should be noted however that, as the changing pattern of heavy fuel oil consumption will reach a new balance only gradually, there will be a shift to light and medium products around 1990.

Table 4:   Structure of consumption of oil products

| | | | | | % |
Products	1980	1983	1990	1995	2000
Gaseous products	3.0	4.1	1.2	1.2	1.1
Light products	25.8	31.9	33.4	31.2	31.9
Medium products	23.5	30.1	31.8	30.2	29.0
Heavy Products	47.7	33.9	33.6	37.4	38.1

The only Irish refinery was purchased from the major oil companies by the national oil company. It currently has a distillation capacity of 2.8 million toe. By contract it has to supply at least 35% of national consumption, in order to divesify Ireland's sources of supply. The increase in consumption of oil products and the shift to light and medium products in the medium term could justify the installation of a cracking plant which would facilitate adjustment to the pattern of demand and make the Irish energy system more flexible.

(b)   Natural gas

12.   Natural gas constitutes a major asset for the policy of
      diversifying energy supplies: its contribution to GIC increased
      from 8.8% in 1980 to nearly 23% in 1983 as national resources came
      on stream. Eventually, as deliveries stabilize, the proportion of
      GIC covered by natural gas will fall to settle at around 15%.   The
      indigenous resources of the offshore Kinsale field amount to
      34.0 million toe and should thus be sufficient to meet the needs of
      the country up to 2000.   Consequently, there are no plans for
      additional import contracts before that date.

      For the moment, as the field has only just begun to produce and the
      transmission and distribution networks are still being set up, the
      use of natural gas is confined chiefly to the electricity sector,
      and to the ammonium and fertilizer industries (0.4 million toe of
      non-energy demand). However, a small proportion of the gas produced
      is already being distributed in the Cork area, and once Cork and
      Dublin are interconnected, natural gas markets are expected to
      develop in industrial sectors and in the residential and tertiary
      sectors, at the expense of the electricity sector.

(c)   Coal products

13.   In solid fuels, Ireland has extensive peat resources which will
      probably be sufficient to maintain a production rate of 1.1 to
      1.4 million toe until the end of the century. Currently, nearly 65%
      of production is used in power stations, the other 35% being sold
      to the residential and tertiary sectors. There is unlikely to be
      any significant change in this consumption pattern over the period
      except a reduction in the share going to the residential sector as
      living standards improve.

      Ireland's coalfields are small and of low quality, so that the
      extra consumption will have to be covered entirely by imports. The
      development of the infrastructure required to handle these
      quantities of coal is already in hand.

14.   Part of this increase in coal consumption is accounted for by the
      power stations.   In 1980, sales of coal to power stations were
      still extremely limited (0.03 million toe), but consumption should
      increase rapidly to reach 1 million toe in 1990 and will contrinue
      to rise thereafter.   In addition, a major campaign is currently in
      progress to encourage conversion to coal in the industrial sectors.
      The result should be an appreciable increase (of 0.4 million toe in
      2000) in deliveries of coal to these sectors.

(d)   Electricity

15.   The main feature of the electricity sector is the growing share of
      coal in meeting demand, which will increase by 2.2% a year between
      1983 and 1990 and 2.6% between 1991 and 2000, i.e. an average
      annual growth rate of 2.4% over the whole period.

The investment programme is directed primarily towards the use of coal   Between 1980 and 2000, an additional 1 385 MWe of fossil-fuelled power stations are to come on stream.

Table 5:   Capital investment programme for the electricity sector

Plant	1981-1990	1991-1995	1996-2000
Conv. power stations	922 MW	-	463 MW
Peak-load power stations	503 MW	-	-
Renewable sources (wood, wind, etc.)	20 MW	27 MW	75 MW

A breakdown of the overall production structure by type of fuel (all producers) shows the very substantial share taken by solid fuels in the medium and long term. It is primarily gaseous fuels and to a lesser extent liquid fuels which will suffer from the competition of solid fuels since their use will gradually be confined to central-load plants.

Table 6: Total net electricity production

TWh and %

Origin	1983 TWh	1983 %	1990 TWh	1990 %	1995 TWh	1995 %	2000 TWh	2000 %
Hydro	1.2	10.9	0.9	7.7	0.9	6.7	0.9	6.0
Solid fuels	1.7	16.1	7.8	66.7	9.3	69.4	11.0	72.8
Liquid fuels	2.3	21.2	1.2	10.3	1.6	11.9	1.3	8.6
Gas. fuels	15.5	51.8	1.7	14.5	1.4	10.5	1.3	8.6
Other	-	-	0.1	0.8	0.2	1.5	0.6	4.0
Total	10.7	100.0	11.7	100.0	13.4	100.0	15.1	100.0

D. SUMMARY TABLES
ENERGY SUPPLY/DEMAND BALANCE - 1980   **IRELAND**

million toe

	SOLID FUEL	PETROL PRODCT	NAT GAS	NUCLR ENERGY	HEAT	HYDRO+ OTHERS	RENEW ENER	TOTAL
PRIM PRODUCT	1.2	-	0.7	-	-	0.1	-	2.0
TOT IMPORTS	0.8	5.9	-	-	-	-	-	6.7
TOT EXPORTS	0.0	0.2	-	-	-	-	-	0.2
STOCK CHANGE	0.0	0.0	-	-	-	-	-	0.0
GROSS CONSUM	2.0	5.7	0.7	-	-	0.1	-	8.5
BUNKERS	-	0.1	-	-	-	-	-	0.1
INLAND CONSM	2.0	5.6	0.7	-	-	0.1	-	8.4
ELEC POWER S	0.6	1.4	0.4	-	-	-0.8	-	1.6
OTHER TRANSF	0.0	0.2	-	-	-	-	-	0.2
ENERG INDUST	-	0.1	-	-	-	0.1	-	0.2
FINAL CONSUM	1.4	3.9	0.3	-	-	0.8	-	6.4
- NON ENERGY	-	0.1	-	-	-	-	-	0.1
- INDUSTRY	0.1	1.2	0.3	-	-	0.3	-	1.9
- TRANSPORT	-	1.7	-	-	-	-	-	1.7
- HOUSEHOLD	1.3	0.9	-	-	-	0.5	-	2.7
STAT DIFFER	0.0	0.0	-	-	-	-	-	0.0

ENERGY SUPPLY / DEMAND BALANCE - 1983

million toe

	SOLID FUEL	PETROL PRODCT	NAT GAS	NUCLR ENERGY	HEAT	HYDRO+ OTHERS	RENEW ENER	TOTAL
PRIM PRODUCT	1.0	-	1.8	-	-	0.1	-	2.9
TOT IMPORTS	1.0	4.4	-	-	-	-	-	5.4
TOT EXPORTS	0.0	0.4	-	-	-	-	-	0.4
STOCK CHANGE	0.0	0.2	-	-	-	-	-	0.2
GROSS CONSUM	2.0	4.2	1.8	-	-	0.1	-	8.1
BUNKERS	-	0.1	-	-	-	-	-	0.1
INLAND CONSM	2.0	4.1	1.8	-	-	0.1	-	8.0
ELEC POWER S	0.6	0.5	1.3	-	-	-0.9	-	1.6
OTHER TRANSF	0.1	-	-	-	-	-	-	0.1
ENERG INDUST	-	-	-	-	-	0.2	-	0.2
FINAL CONSUM	1.3	3.6	0.5	-	-	0.8	-	6.1
- NON ENERGY	-	0.1	0.4	-	-	-	-	0.4
- INDUSTRY	0.2	1.0	0.0	-	-	0.3	-	1.6
- TRANSPORT	-	1.6	-	-	-	-	-	1.6
- HOUSEHOLD	1.1	0.6	0.1	-	-	0.5	-	2.2
STAT DIFFER	-	0.3	-	-	-	-	-	0.3

ENERGY SUPPLY / DEMAND BALANCE - 1990   **IRELAND**

million toe

	SOLID FUEL	PETROL PRODCT	NAT GAS	NUCLR ENERGY	HEAT	HYDRO+ OTHERS	RENEW ENER	TOTAL
PRIM PRODUCT	1.2	–	1.4	–	–	0.1	0.1	2.8
TOT IMPORTS	2.1	5.3	–	–	–	–	–	7.4
TOT EXPORTS	–	–	–	–	–	–	–	–
STOCK CHANGE	–	–	–	–	–	–	–	–
GROSS CONSUM	3.3	5.3	1.4	–	–	0.1	0.1	10.2
BUNKERS	–	0.1	–	–	–	–	–	0.1
INLAND CONSM	3.3	5.2	1.4	–	–	0.1	0.1	10.1
ELEC POWER S	1.8	0.3	0.4	–	–	-0.8	0.1	1.8
OTHER TRANSF	–	0.1	–	–	–	–	–	0.1
ENERG INDUST	–	0.1	–	–	–	0.1	–	0.2
FINAL CONSUM	1.5	4.7	1.0	–	–	0.9	–	8.1
– NON ENERGY	0	0.2	0.4	–	–	–	–	0.6
– INDUSTRY	0.4	1.5	0.3	–	–	0.3	–	2.5
– TRANSPORT	–	2.0	–	–	–	–	–	2.0
– HOUSEHOLD	1.1	1.0	0.3	–	–	0.6	–	3.0

ENERGY SUPPLY / DEMAND BALANCE - 1995

million toe

	SOLID FUEL	PETROL PRODCT	NAT GAS	NUCLR ENERGY	HEAT	HYDRO+ OTHERS	RENEW ENER	TOTAL
PRIM PRODUCT	1.3	–	1.6	–	–	0.1	0.1	3.1
TOT IMPORTS	2.4	5.7	–	–	–	–	–	8.1
TOT EXPORTS	–	–	–	–	–	–	–	–
STOCK CHANGE	–	–	–	–	–	–	–	–
GROSS CONSUM	3.7	5.7	1.6	–	–	0.1	0.1	11.2
BUNKERS	–	0.1	–	–	–	–	–	0.1
INLAND CONSM	3.7	5.6	1.6	–	–	0.1	0.1	11.1
ELEC POWER S	2.3	0.4	0.3	–	–	-1.0	0.1	2.1
OTHER TRANSF	–	0.0	–	–	–	–	–	0.0
ENERG INDUST	–	0.1	–	–	–	0.1	–	0.2
FINAL CONSUM	1.4	5.1	1.3	–	–	1.0	–	8.8
– NON ENERGY	0	0.2	0.4	–	–	–	–	0.6
– INDUSTRY	0.4	1.7	0.4	–	–	0.4	–	2.9
– TRANSPORT	–	2.2	–	–	–	–	–	2.2
– HOUSEHOLD	1.0	1.0	0.5	–	–	0.6	–	3.1

ENERGY SUPPLY / DEMAND BALANCE - 2000   **IRELAND**

million toe

	SOLID FUEL	PETROL PRODCT	NAT GAS	NUCLR ENERGY	HEAT	HYDRO+ OTHERS	RENEW ENER	TOTAL
PRIM PRODUCT	1.4	-	1.7	-	-	0.1	0.2	3.4
TOT IMPORTS	2.9	5.9	-	-	-	-	-	8.8
TOT EXPORTS	-	-	-	-	-	-	-	-
STOCK CHANGE	-	-	-	-	-	-	-	-
GROSS CONSUM	4.3	5.9	1.7	-	-	0.1	0.2	12.2
BUNKERS	-	0.1	-	-	-	-	-	0.1
INLAND CONSM	4.3	5.8	1.7	-	-	0.1	0.2	12.1
ELEC POWER S	2.7	0.3	0.3	-	-	-1.2	0.2	2.3
OTHER TRANSF	-	0.0	-	-	-	-	-	0.0
ENERG INDUST	-	0.1	-	-	-	0.2	-	0.3
FINAL CONSUM	1.6	5.4	1.4	-	-	1.1	-	9.5
- NON ENERGY	0	0.2	0.4	-	-	-	-	0.6
- INDUSTRY	0.6	1.6	0.5	-	-	0.4	-	3.2
- TRANSPORT	-	2.4	-	-	-	-	-	2.4
- HOUSEHOLD	1.0	1.1	0.5	-	-	0.7	-	3.3

Summarized Energy Balance - IRELAND

in million toe	1973[a]	1980[a]	1983[a]	1990[b]	1995[b]	2000[b]
I. Gross Energy Consumption	7.02	8.50	8.02	10.20	11.20	12.20
- Bunkers	0.07	0.07	0.05	0.10	0.10	0.10
- Inland consumption	6.95	8.43	7.97	10.10	11.10	12.10
II. Inland Energy Consumption	6.95	8.43	7.97	10.10	11.10	12.10
- Solid fuels	1.39	1.99	1.91	3.30	3.70	4.30
- Oil	5.50	5.63	4.21	5.20	5.60	5.80
- Gas	-	0.74	1.78	1.40	1.60	1.70
- Primary electricity, etc.	0.06	0.07	0.07	0.20	0.20	0.30
III. Indigenous Production[1]	0.74	1.97	2.88	2.80	3.10	3.40
- Hard coal	0.03	0.03	0.04	0.00	-	-
- Lignite & peat	0.65	1.13	1.00	1.20	1.30	1.40
- Oil	-	-	-	-	-	-
- Natural gas	-	0.74	1.78	1.40	1.60	1.70
- Nuclear energy	-	-	-	-	-	-
- Hydro & geothermal[2]	0.06	0.07	0.07	0.10	0.10	0.10
- Others & renewables	-	-	-	0.10	0.10	0.20
IV. Net Imports[3]	6.02	6.53	4.96	7.40	8.10	8.80
- Solid fuels	0.49	0.79	0.94	2.10	2.40	2.90
- Oil	5.52	5.74	4.02	5.30	5.70	5.90
- Natural gas[2]	-	-	-	-	-	-
- Electricity[2]	0.01	-	-	-	-	-
V. Stock Changes[4]	-0.27	0.00	-0.17	-	-	-
- Solid fuels	-0.22	-0.04	0.07	-	-	-
- Oil	-0.05	0.04	-0.24	-	-	-
- Gas	-	-	-	-	-	-
VI. Electricity Generation Input	1.84	2.49	2.56	2.70	3.20	3.60
- Solid fuels[5]	0.63	0.63	0.63	1.80	2.30	2.70
- Oil	1.15	1.41	0.55	0.30	0.40	0.30
- Natural gas	-	0.38	1.31	0.40	0.30	0.30
- Nuclear energy	-	-	-	-	-	-
- Hydro & geothermal[2]	0.06	0.07	0.07	0.10	0.10	0.10
- Others & renewables	-	-	-	0.10	0.10	0.20

Main indicators (related to long term objectives)

	1973-1963	1980-1975	1983-1980	1990-1983	2000-1990
Inland energy annual growth rates	+6.5%	+5.8%	-1.9%	+3.4%	+1.8%
GDP annual growth rates	+4.4%	+4.4%	+1.8%	+3.7%	+3.4%
Energy-GDP ratio	1.48	1.32	-1.06	0.92	0.53

	1973	1980	1983	1990	1995	2000
Share of oil in gross energy consumption	79.3%	67.1%	53.1%	51.9%	51.0%	48.4%
Share of coal and nuclear in in electricity production	34.2%	25.3%	24.6%	66.7%	71.9%	75.0%
Supply dependance on imports	85.8%	76.8%	61.8%	72.5%	72.3%	72.1%

Sources: a. Statistical Office of the European Communities
        b. "Energy 2000"

Notes: 1. Production of primary sources, including recovered products.
      2. The conversion of electricity, including hydro and geothermal, is
         based on its actual energy content: 3600 kjoules/kWh or 860 kcal/kWh
      3. The (-) sign means net exports
      4. The (-) sign means a stock decrease
      5. Including coke oven gas and blast furnace gas (derived from coal)

I T A L Y

A. ECONOMIC REFERENCE FRAMEWORK

1. For the period 1983 to 2000, the GDP growth assumption is ·an average annual rate of 2.8%, the same in both periods, 1983-90 and 1990-2000.

From 1983 to 1990 public-sector consumption is expected to increase at an average rate of 1.6% per annum and public-sector investment at 5.7% p.a. The real increase in earnings (1% per annum) will stimulate private consumption, which will grow at an average annual rate of 2.3%. Unlike the trend in the other countries, gross fixed capital formation will increase more slowly (1.7%) than private consumption, owing in particular to high real interest rates on the public debt. In a more favourable international context, there will be strong export growth (5% per annum), which will benefit such industries as capital goods, food and textiles.

The trends over the period up to 1990 will continue broadly unchanged from 1991 to 2000, but probably with stronger growth of gross fixed capital formation, which will promote activity in the capital goods and building industries.

Sector by sector, the assumptions adopted point to the following results:

- consumer goods industries such as food processing and textiles benefit from buoyant private consumption and from expanding exports;

- building and non-metal mineral products record indifferent results as a consequence of several contradictory factors: relatively low investment in manufacturing (at least until 1990), the moderate level of housebuilding (increasing by 1.7% per annum), and buoyant public investment; after 1990, the results will be better as investment recovers;

- the capital goods industries experience fairly strong growth until the end of the period, benefiting from the export trend and renewed investment after 1990;

- the service sector's share of GDP continues to rise, from 52% in 1983 to 54% in 2000.

Overall, growth will be led mainly by the service and capital goods sectors, which accounts for their growing shares in the value added generated by the productive system (see Table 1).

Table 1: Sectoral trends

LIT '000 million (1975)

	1970	AAGR[24]	1980	AAGR	1983	AAGR	1990	AAGR	2000
Agriculture	9.1	1.3	10.4	0.3	10.5	1.6	11.0	1.0	12.2
Industry	38.6	3.8	55.9	-2.2	52.3	3.1	64.7	2.6	83.4
EIIs[25]	10.3	4.4	15.8	-1.7	15.0	1.6	16.8	1.9	20.3
Other industries	28.3	3.5	40.1	-2.4	37.3	3.6	47.9	2.8	63.1
Building	10.7	0.0	10.7	-0.9	10.4	1.8	11.8	3.0	15.9
Services	55.1	3.5	77.4	0.6	78.9	3.0	97.1	3.0	130.5
Total value added (GDP)	113.5	3.1	154.4	-0.5	152.1	2.8	185.6	2.8	242.0

B. **TOTAL FINAL ENERGY DEMAND**[26]

2. Total final energy demand between now and the year 2000 will
    continue to be largely determined by economic activity. Energy
    demand will increase at an average rate of 1.4% per annum compared
    with a 2.8% economic growth rate; it will be led by industrial
    consumption (+1.7% p.a.) and underpinned by consumption in the
    residential and tertiary sector (+1.3% p.a.) and the transport
    sector (+1% p.a.).

    The structure of the energy balance will show certain changes:
    gas will take over much of the share of oil products in heating
    uses, and to a lesser extent will replace electricity, where
    specific requirements will increase faster than the other energy
    uses. Coal will have a smaller but still important role, as it
    will continue to be required for certain industrial uses. New
    energy sources will appear in the energy balance in 2000, but
    their role will remain marginal.

a) **Industry**

3. The growth of energy demand in the industrial sector will continue
    to be linked with industrial growth: 1.7% per annum on average
    between now and 2000 against 2.7% p.a. for total value added of
    industries.

    Use of more energy-thrifty processes and the general adoption of
    energy-saving measures will reduce the energy intensity per unit
    of value added of the industrial sector (average gain of 25%): it
    will fall from 665 (10^3 toe/10^{12} LIT (1975)) to 550 in 1990 and
    496 in 2000.

---

[24] AAGR: average annual growth rate.
[25] EEI: energy-intensive industries = metals + non-metal minerals + paper
+ chemicals.
[26] The trend in total final energy demand for Italy was determined using
MEDEE 3 for the residential and tertiary sector and a simplified
version (MEDEE 2) for the industrial sector. The transport sector was
the subject of a non-model evaluation.

In 1983 the energy-intensive industries accounted for 75% of the energy demand of industry and their consumption continues to grow at the rate of 1.1% p.a. while their activity expands at 1.8% p.a: this largely explains how industrial activity continues to stimulate energy demand.

A further spur is the vigorous increase in consumption by light industry (2.5% p.a.), certainly sustained by buoyant economic activity, especially in the capital goods sector.

The breakdown of consumption by end-uses will not change significantly, except that specific electricity requirements will rise somewhat faster than the average (2.6% p.a.). In fact the increase in the demand for electricity will account for much of the general increase in energy demand between now and 2000, the rest being met mainly by natural gas, with little change in the share of oil products.

b)   Transport

4. Between 1983 and 1990, energy consumption in the transport sector will increase by 1.5% per annum, compared with average GDP growth of 2.8%; from 1990 to 2000 the uncoupling of the growth of energy consumption in the transport sector (1.0% per annum) from GDP growth (2.8%/year) is more marked.

As a result, unit energy consumption per unit of GDP will fall by almost 20% between 1983 and 2000, from 0.167 toe per million LIT in 1983 to 0.160 in 1990 and 0.135 in 2000.

This trend, all the more remarkable for the fact that vehicle ownership is likely to increase faster than in most other Community countries, will stem from a considerable improvement in vehicle efficiency.

5. Motor fuels will continue to account for approximately 98% of consumption in the transport sector until the end of the century. As in the other Community countries, petrol will yield ground to automotive gas oil, possibly quite fast, to judge from recent observations.

c)   Residential and tertiary sector

6. Energy consumption in the residential and tertiary sector is expected to grow fairly slowly (at 1.3% per year), but faster than in many other European countries. In the residential sector as in the tertiary sector, growth will be led mainly by the increase in energy end-uses not specifically connected with heating. These, however, account for only 30% of consumption in the residential sector and 44% of consumption in the tertiary sector, heating taking up the remainder.

7. The structure of the housing stock will remain relatively stable, both in the age of buildings and in the breakdown between apartment blocks and one-family houses: only 23% of housing units in 2000 will be less than 25 years old, and apartments will still make up approximately 80% of the total. This lack of change means that the binding energy-saving regulations applying to new dwellings will affect only a small proportion of the total. What is more, these efforts will be offset by the increase in central heating as a result of new construction. There will be greater insulation of older housing, however, which will produce energy savings of 5-8% depending on the type of dwelling and the region. All in all, unit energy consumption for heating would decrease on average by 3% by 2000. As the total housing stock will increase by 8%, the net result will be an average annual increase of 0.5% p.a. in total energy demand for heating purposes between now and 2000.

Energy demand for the other end-uses will grow more quickly, especially for domestic hot water (2.8% p.a.) and specific electricity (2.3% p.a.).

8. A faster replacement rate for buildings in the tertiary sector than for housing, accompanied by radical energy-saving measures in new buildings, will reduce heating requirements in the tertiary sector by 11%, despite a 15% increase in the total volume of premises.

The situation is quite different for other end-uses and in particular specific electricity, including air-conditioning, which will increase at an average rate of 2.8% pa.

9. A marked shift from gas oil to natural gas significantly modifies the final energy balance of the residential and tertiary sector: from 41% of demand in 1980, the share of gas oil falls to 29% in 2000 while that of gas goes up from 25% to 39%.

Less significant changes are caused by an expansion in district heating networks, which will account for 4% of demand in 2000.

The share of electricity will rise (21% in 2000 against 17% at present), boosted by the increase in specific electricity requirements and, to a lesser extent, demand for domestic hot water. Consumption for heating purposes will remain quite marginal.

Overall, the energy intensity of the residential and tertiary sector will fall from 223 (10^3 toe/10^{12} LIT (1975)) in 1983 to 203 in 1990 and 174 in 2000, a drop of 22% in 17 years.

Table 2: Final energy consumption

ITALY	million toe			Annual % changes	
	1983	1990	2000	1990/83	2000/90
Industry	30.9	35.6	41.4	+2.0%	+1.5%
Transport	25.4	29.5	32.6	+2.2%	+1.0%
Residential and					
tertiary	33.9	37.5	42.1	+1.5%	+1.2%
Non-energy uses	8.4	8.7	9.4	+0.5%	+0.8%
Total	98.6	111.3	125.5	+1.7%	+1.2%
of which					
Solid fuels	4.7	5.2	6.1	+1.5%	+1.6%
Oil products	57.9	62.0	63.1	+1.0%	+0.2%
Gas	20.5	27.0	33.5	+3.9%	+2.2%
Electricity	13.8	16.5	21.0	+2.6%	+2.4%
Other	-	0.6	1.8	..	+11.6%

C. ENERGY SUPPLY

10. Table 3, a breakdown of gross inland primary energy consumption
(GIC) based on the economic scenario adopted, shows an increase in
GIC of about 2.1% p.a. up to 1990 and 1.6% p.a. thereafter, making
an average annual growth rate of 1.8% over the period 1983-2000.

Table 3: Gross inland primary energy consumption

million toe

	1983	1990	1995	2000
Coal products	12.8	21.5	25.8	31.0
Oil products	83.1	79.5	76.9	77.4
Natural gas	22.4	32.7	32.7	35.1
Nuclear fuels	1.6	5.4	14.6	19.4
Primary electricity	5.1	6.0	6.1	5.9
Renewable energy sources	0.2	0.2	0.2	0.3
GIC	125.2	145.1	157.3	169.1

The structure of Italy's energy supply continues to be dominated by
oil products, although their share declines from 65% in 1983 to 55%
in 1990 and 46% in 2000. The shift for final consumers is to solid
fuels and natural gas and for electricity generation to coal and
nuclear energy.

As Italy pursues its nuclear programme in the medium term and
continues its exploitation of oil and gas at present levels, its
dependence on imported energy will fall from 83.6% in 1983 to 76.3%
in 2000.

a) Oil products

11. Despite their declining share in GIC, consumption of oil products
remains fairly stable throughout the entire period. This apparent
stability is the net result of several trends:

- a sharp drop in the consumption of extra-heavy fuel oil in power stations (12.7 million toe in 1990 and 9.5 million toe in 2000, against 19.6 million toe in 1983), as the conversion of oil-fired power stations to coal continues;

- fairly steady sales of heavy fuel oil to industry (9.2 million toe in 2000 compared with 8.1 million toe in 1983);

- only a slight fall in heating oil sales to the residential and tertiary sectors (10.5 million toe in 1990 and 9.8 million toe in 2000 against 13.6 million toe in 1983) despite strong competition from natural gas, distributed heat and electricity;

- increased consumption of automotive gas oil in the transport sector, from 9.9 million toe in 1983 to 12.1 million toe in 1990 and 13.5 million toe in 2000, with the expansion in road haulage and the growing shift to diesel fuel for private cars;

- steady sales to the petrochemical industry;

- increased petrol sales, from 12.1 million toe in 1983 to 12.8 million toe in 1990 and 13.9 million toe in 2000, due to the increase in total motor fuel demand by private individuals, which makes up for the market shares lost to automotive gas oil.

12.   As total consumption of oil products remains steady, the drop in sales of extra heavy fuel oil to power stations means a sharp increase in the share of light and medium products in the structure of consumption (Table 4).

Table 4:   Structure of consumption of oil products

Products	1980	1983	1990	1995	2000
Gaseous products	3.2	3.2	3.4	3.6	3.6
Light products	21.5	21.8	25.1	26.5	26.7
Medium products	26.7	29.6	30.1	31.6	32.4
Heavy products	48.6	45.4	41.4	38.3	37.3

Given Italy's refining capacity in service at 1 January 1984 - 152.6 million tonnes of atmospheric distillation capacity, 13.6 million tonnes of catalytic cracking and 14.2 million tonnes of visbreaking - the refining sector, which is dominated by simple topping-reforming refineries, has substantial excess capacity. This is nothing new: 1973 was the last year in which the rate of utilization of distillation capacities exceeded 65% (52% in 1983). As the volume of crude oil processed in refineries will remain at steady levels, refining capacity must be rationalized to return it to profitability.

With the steady decline in sales of heavy fuel oil, restructuring could focus on the complex refineries which now account for nearly 90 million tonnes of capacity. If this is done, additional investment in conversion units to meet structural changes in demand will be minimal since so much will already have been done. However, some simple refineries will continue to operate in regions with significant outlets for heavy fuel oil (industrial regions) since the location of this type of refinery depends directly on the existence of such markets.

b)   Natural gas

13.   The share of natural gas in GIC rises substantially from 17.5% in 1983 to 22% in 1990 and remains steady thereafter.   This marked increase comes about as the new USSR and Algerian contracts take effect.

After 1990 all existing contracts will be in full operation, precluding the need for any major new contracts before the end of the decade. Gas imports from the USSR will account for about 46% of total natural gas imports in 1990 and 48% in 2000. The exploitation of national resources is expected to hold steady at 10 million toe until the end of the century.

14.   The results show that   as transport and distribution networks are developed,   the residential and tertiary sector will provide an ever growing market for natural gas: 12.9 million toe in 1990 and 16.6 million toe in 2000 against 9.5 million toe in 1983.   Industry will increase its total energy demand and take larger deliveries of natural gas: 11.2 million toe in 1990 and 13.3 million toe in 2000 against 9.1 million toe in 1983.   An important factor is Italy's geography: a major transport network is needed for the movement of supplies between places of importation and the consuming areas, and this takes time to build.   For this reason interruptible sales to industry in general and the electricity sector in particular will have an important role to play as market regulator in the years to come.   Sales to the electricity sector will peak at 5.7 million toe in 1990 and then decline throughout the 1990s.

c)   Solid fuels

15.   In the medium term,   consumption of coal products increases considerably,   from 12.8 million toe in 1983 to 21.5 million toe in 1990 and 31.0 million toe in 2000,   their share in GIC rising from 10.0% in 1983 to 14.8% in 1990 and 18.3% in 2000.

As indigenous resources are extremely limited,   the increase in consumption between 1983 and 2000 will mean a corresponding increase in imports and in turn the necessary development of infrastructure (ports, storage areas, transport facilities).

This increase in consumption is due to the gradual conversion of power stations from oil to coal,   so that the electricity sector will   be   substantially   increasing   its   demand   for   coal: 13.4 million toe in 1990 and 28.4 million toe in 2000 against 4.8 million toe in 1983.   Industry will also considerably increase its consumption of steam coal: 1.7 million toe in 1990 and 3.2 million toe in 2000 against 1.4 million toe in 1983.

d)   Electricity

16.   The main feature of the electricity sector is the increasing share
       of nuclear power and coal in meeting demand, set to grow at a rate
       of 2.8% p.a. between 1983 and 1990 and 2.4% between 1991 and 2000,
       making an average annual growth rate of 2.6% over the whole period.

       Medium- and long-term capital investment policy rests on the twin
       pillars of conversion to coal of the oil-fired power stations in
       operation or being built and the development of nuclear energy. Any
       further delay in carrying out this programme (and there have
       already been many delays) would have serious repercussions on
       generating costs and therefore the competitiveness of electricity.
       The effects would of course carry through to the primary energy
       balance and Italy's dependence on imports.

       Table 5:   Capital investment programme for the electricity sector

Plant	1981-1990	1991-1995	1996-2000
Nuclear PWR	2870 Mwe	6000 MWe	3260 MWe
Coal-fired	3528 MWe	900 MWe	3645 MWe
Dual-fired oil/gas	10 880 MWe	–	–
Hydro	5 226 MWe	1 235 MWe	202 MWe
Renewables[27]	75 MWe	76 MWe	65 MWe
CHP	1 590 MWth	3 236 MWth	1 157 MWth

17.   The overall production structure (all producers) by type of fuel
       shows the large shares contributed by nuclear power (27% in 2000)
       and coal (35%). In the long term the share of liquid and gaseous
       fuels shows a steady decline, from 58% in 1982 to 39% in 1990, 24%
       in 1995 and 17% in 2000. Imported electricity, mostly from France,
       continues to account for a significant share until after 1990.

       Table 6:   Total net electricity production

| | | | | | | | TWh and % |
| | 1983 | | 1990 | | 1995 | | 2000 | |
Origin	Twh	%	Twh	%	Twh	%	Twh	%
Hydro	43.9	25.2	43.8	21.2	48.1	19.8	50.0	18.3
Nuclear power	5.6	3.2	20.1	9.7	55.9	23.1	74.2	27.3
Solid fuels	20.2	11.5	59.1	28.6	75.9	31.3	97.3	35.7
Liquid fuels	86.2	49.4	54.2	26.2	41.6	17.2	40.9	15.0
Gaseous fuels	15.4	8.8	26.4	12.8	17.1	7.1	6.0	2.2
Other	3.2	1.9	3.2	1.5	3.6	1.5	4.0	1.5
TOTAL	174.5	100.0	206.9	100.0	242.2	100.0	272.4	100.0

---

[27] Wind and geothermal.

D. SUMMARY TABLES

ENERGY SUPPLY / DEMAND BALANCE - 1980   **ITALY**

million toe

	SOLID FUEL	PETROL PRODCT	GAS	NUCLR ENERGY	HEAT	HYDRO+ OTHERS	RENEW ENER	TOTAL
PRIM PRODUCT	0.3	2.0	10.2	0.7	–	4.1	0.2	17.5
TOT IMPORTS	11.3	107.8	11.8	–	–	0.7	–	131.6
TOT EXPORTS	0.5	12.0	–	–	–	0.2	–	12.7
STOCK CHANGE	-0.1	-0.8	0.7	–	–	–	–	0.2
GROSS CONSUM	11.0	97.0	22.7	0.7	–	4.6	0.2	136.1
BUNKERS	–	4.1	–	–	–	–	–	4.1
INLAND CONSM	11.0	92.9	22.7	0.7	–	4.6	0.2	132.0
ELEC POWER S	3.3	22.5	2.7	0.7	–	-11.7	0.2	17.7
OTHER TRANSF	3.6	1.2	-2.8	–	–	–	–	2.0
ENERG INDUST	0.1	4.9	1.4	–	–	2.5	–	8.9
FINAL CONSUM	4.2	65.4	21.1	–	–	13.8	–	104.5
- NON ENERGY	0.3	6.5	2.1	–	–	–	–	8.9
- INDUSTRY	3.7	15.6	9.8	–	–	8.1	–	37.2
- TRANSPORT	0.0	24.0	0.3	–	–	0.4	–	24.7
- HOUSEHOLD	0.2	19.3	8.9	–	–	5.3	–	33.7
STAT DIFFER	-0.2	-1.1	0.3	–	–	–	–	-1.0

ENERGY SUPPLY / DEMAND BALANCE - 1983

million toe

	SOLID FUEL	PETROL PRODCT	GAS	NUCLR ENERGY	HEAT	HYDRO+ OTHERS	RENEW ENER	TOTAL
PRIM PRODUCT	0.3	2.2	10.6	1.6	–	4.2	0.2	19.1
TOT IMPORTS	12.3	92.4	12.1	–	–	1.2	–	118.0
TOT EXPORTS	0.1	12.8	–	–	–	0.3	–	13.2
STOCK CHANGE	0.3	4.5	-0.3	–	–	–	–	4.5
GROSS CONSUM	12.8	86.3	22.4	1.6	–	5.1	0.2	128.4
BUNKERS	–	3.2	–	–	–	–	–	3.2
INLAND CONSM	12.8	83.1	22.4	1.6	–	5.1	0.2	125.2
ELEC POWER S	4.8	19.6	3.6	1.6	–	-11.3	0.2	18.5
OTHER TRANSF	3.3	0.7	-2.6	–	–	–	–	1.5
ENERG INDUST	–	4.9	0.9	–	–	2.6	–	8.4
FINAL CONSUM	4.7	57.9	20.5	–	–	13.8	–	96.9
- NON ENERGY	0.2	6.6	1.5	–	–	–	–	8.4
- INDUSTRY	4.1	10.4	8.7	–	–	7.4	–	30.9
- TRANSPORT	–	24.8	0.3	–	–	0.4	–	25.4
- HOUSEHOLD	0.3	17.7	9.9	–	–	6.0	–	33.9
STAT. DIFFER	–	-1.6	0.1	–	–	–	–	-1.5

ENERGY SUPPLY / DEMAND BALANCE - 1990   **ITALY**

million toe

	SOLID FUEL	PETROL PRODCT	GAS	NUCLR ENERGY	HEAT	HYDRO+ OTHERS	RENEW ENER	TOTAL
PRIM PRODUCT	1.5	4.0	10.0	5.4	-	4.8	0.2	25.9
TOT IMPORTS	20.4	86.7	22.7	-	-	1.4	-	131.2
TOT EXPORTS	0.4	7.0	-	-	-	0.2	-	7.6
STOCK CHANGE	-	-	-	-	-	-	-	-
GROSS CONSUM	21.5	83.7	32.7	5.4	-	6.0	0.2	149.5
BUNKERS	-	4.2	-	-	-	-	-	4.2
INLAND CONSM	21.5	79.5	32.7	5.4	-	6.0	0.2	145.3
ELEC POWER S	13.4	12.7	6.0	5.4	-0.6	-13.7	0.2	23.4
OTHER TRANSF	2.9	0.6	-1.5	-	-	-	-	2.0
ENERG INDUST	-	4.2	1.2	-	-	3.2	-	8.6
FINAL CONSUM	5.2	62.0	27.0	-	0.6	16.5	0.0	111.3
- NON ENERGY	0.3	5.8	2.6	-	-	-	-	8.7
- INDUSTRY	4.5	10.5	11.2	-	0.2	9.2	-	35.6
- TRANSPORT	-	28.8	0.3	-	-	0.4	-	29.5
- HOUSEHOLD	0.4	16.9	12.9	-	0.4	6.9	0.0	37.5

ENERGY SUPPLY / DEMAND BALANCE - 1995

million toe

	SOLID FUEL	PETROL PRODCT	GAS	NUCLR ENERGY	HEAT	HYDRO+ OTHERS	RENEW ENER	TOTAL
PRIM PRODUCT	1.5	4.0	10.0	14.6	-	5.2	0.2	35.5
TOT IMPORTS	24.7	84.1	23.7	-	-	1.1	-	133.6
TOT EXPORTS	0.4	7.0	-	-	-	0.2	-	7.6
STOCK CHANGE	-	-	-	-	-	-	-	-
GROSS CONSUM	25.8	81.1	33.7	14.6	-	6.1	0.2	161.5
BUNKERS	-	4.2	-	-	-	-	-	4.2
INLAND CONSM	25.8	76.9	33.7	14.6	-	6.1	0.2	157.3
ELEC POWER S	17.7	9.8	3.9	14.6	-1.3	-16.4	0.2	28.6
OTHER TRANSF	2.6	0.6	-1.4	-	-	-	-	1.8
ENERG INDUST	-	4.2	1.3	-	-	3.6	-	9.1
FINAL CONSUM	5.5	62.3	29.9	-	1.3	18.9	0.0	117.9
- NON ENERGY	0.3	5.8	3.0	-	-	-	-	9.1
- INDUSTRY	4.9	10.5	12.0	-	0.3	10.3	-	38.0
- TRANSPORT	-	30.4	0.3	-	-	0.5	-	31.2
- HOUSEHOLD	0.3	15.6	14.6	-	1.0	8.1	0.0	39.6

ENERGY SUPPLY / DEMAND BALANCE - 2000   **ITALY**

million toe

	SOLID FUEL	PETROL PRODCT	GAS	NUCLR ENERGY	HEAT	HYDRO+ OTHERS	RENEW ENER	TOTAL
PRIM PRODUCT	1.5	4.5	10.0	19.4	–	5.5	0.3	41.2
TOT IMPORTS	29.9	84.1	25.1	–	–	0.6	–	139.7
TOT EXPORTS	0.4	7.0	–	–	–	0.2	–	7.6
STOCK CHANGE	–	–	–	–	–	–	–	–
GROSS CONSUM	31.0	81.6	35.1	19.4	–	5.9	0.3	173.3
BUNKERS	–	4.2	–	–	–	–	–	4.2
INLAND CONSM	31.0	77.4	35.1	19.4	–	5.9	0.3	169.1
ELEC POWER S	22.4	9.5	1.5	19.4	-1.8	-19.1	0.3	32.2
OTHER TRANSF	2.5	0.6	-1.3	–	–	–	–	1.8
ENERG INDUST	–	4.2	1.4	–	–	4.0	–	9.6
FINAL CONSUM	6.1	63.1	33.5	–	1.8	21.0	0.0	125.5
- NON ENERGY	0.3	5.8	3.3	–	–	–	–	9.4
- INDUSTRY	5.5	10.6	13.3	–	0.5	11.5	–	41.4
- TRANSPORT	–	31.7	0.3	–	–	0.6	–	32.6
- HOUSEHOLD	0.3	15.0	16.6	–	1.3	8.9	0.0	42.1

Summarized Energy Balance – ITALY

in million toe	1973[a]	1980[a]	1983[a]	1990[b]	1995[b]	2000[b]
I. Gross Energy Consumption	129.24	136.10	130.05	149.5	161.5	173.3
– Bunkers	7.03	4.13	3.17	4.2	4.2	4.2
– Inland consumption	122.21	131.97	126.88	145.3	157.3	169.1
II. Inland Energy Consumption	122.21	131.97	126.88	145.3	157.3	169.1
– Solid fuels	8.08	10.91	12.78	21.5	25.8	31.0
– Oil	95.20	92.87	83.15	79.5	76.9	77.4
– Gas	14.23	22.72	22.53	32.7	33.7	35.1
– Primary electricity, etc.	4.70	5.47	8.42	11.6	20.9	25.6
III. Indigenous Production[1]	19.33	17.51	20.76	25.9	35.5	41.2
– Hard coal	0.00	–	–	–	–	–
– Lignite & peat	0.31	0.31	0.29	1.5	1.5	1.5
– Oil	1.77	1.99	2.24	4.0	4.0	4.5
– Natural gas	12.62	10.26	10.76	10.0	10.0	10.0
– Nuclear energy	0.93	0.67	1.63	5.4	14.6	19.4
– Hydro & geothermal[2]	3.44	4.12	5.68	4.8	5.2	5.5
– Others & renewables	0.26	0.16	0.16	0.2	0.2	0.3
IV. Net Imports[3]	112.09	118.79	104.80	123.6	126.0	132.1
– Solid fuels	7.71	10.75	12.16	20.0	24.3	29.5
– Oil	102.65	95.76	79.59	79.7	77.1	77.1
– Natural gas[2]	1.65	11.76	12.10	22.7	23.7	25.1
– Electricity[4]	0.08	0.52	0.95	1.2	0.9	0.4
V. Stock Changes[4]	2.18	0.20	-4.49	–	–	–
– Solid fuels	-0.06	0.15	-0.33	–	–	–
– Oil	2.20	0.75	-4.49	–	–	–
– Gas	0.04	-0.70	0.33	–	–	–
VI. Electricity Generation Input	26.07	33.57	35.46	42.5	51.6	58.5
– Solid fuels[5]	1.29	4.19	5.46	14.6	18.8	23.4
– Oil	19.13	22.47	19.60	12.7	9.8	9.5
– Natural gas	1.02	1.96	2.93	4.8	2.8	0.4
– Nuclear energy	0.93	0.67	1.63	5.4	14.6	19.4
– Hydro & geothermal[2]	3.44	4.12	5.68	4.8	5.2	5.5
– Others & renewables	0.26	0.16	0.16	0.2	0.2	0.3

Main indicators (related to long term objectives)

	1973–1963	1980–1975	1983–1980	1990–1983	2000–1990
Inland energy annual growth rates	+7.2%	+1.9%	-1.3%	+1.9%	+1.5%
GDP annual growth rates	+4.9%	+3.8%	-0.5%	+2.8%	+2.8%
Energy–GDP ratio	1.47	0.50	0.38	0.70	0.54

	1973	1980	1983	1990	1995	2000
Share of oil in gross energy consumption	79.1%	71.3%	66.4%	56.0%	50.2%	47.1%
Share of coal and nuclear in in electricity production	0.9%	14.5%	20.0%	47.1%	64.7%	73.2%
Supply dependance on imports	86.7%	87.3%	80.6%	82.7%	78.0%	76.2%

Sources: a. Statistical Office of the European Communities
          b. "Energy 2000"
Notes: 1. Production of primary sources, including recovered products.
         2. The conversion of electricity, including hydro and geothermal, is
            based on its actual energy content: 3600 kjoules/kWh or 860 kcal/kWh
         3. The (–) sign means net exports
         4. The (–) sign means a stock decrease
         5. Including coke oven gas and blast furnace gas (derived from coal)

L U X E M B O U R G

A. ECONOMIC REFERENCE FRAMEWORK

1. Før the period 1983 to 2000, the GDP growth assumption for Luxembourg is an average annual rate of 2.2%, spread unevenly over the period: 2.0% p.a. in the period 1983-90 and 2.3% p.a. in the following decade.

After several years of slower GDP growth, a reversal in the trend is expected in 1984-85 which should continue throughout the 1980s. In a more favourable world economic climate, Luxembourg exports, although held back by limited steel sector prospects, will benefit from the policy for industrial diversification. Private consumption will reflect the limited rise in wages and salaries. Investment, especially for restructuring, will follow a favourable trend in the steel sector and in the small traditional industries, but investment in housebuilding will be restricted by the unfavourable trend in household incomes.

The trends observed over the period 1983-90 will continue over the period 1991-2000 but with a much more optimistic outlook. The average GDP growth rate will approach 2.3% p.a. under the impetus of the substantial development of the services sector.

Sector by sector, the assumptions adopted point to the following results:

- industry becomes increasingly diversified, thus shaking off its constraining reliance on a single product (steel) and achieving moderate growth;

- the building and public works sector faces a doubtful outlook, both in housebuilding and in public investment;

- the services sector, especially banking, will continue to increase its share in GDP, from 56% in 1983 to 63% in 2000.

Overall, growth will be led mainly by the services sector, as Table 1 shows.

Table 1: Sectoral trends

LFR million at 1975 prices	1970	AAGR[28]	1980	AAGR	1983	AAGR	1990	AAGR	2000
Agriculture	2749	0.6	2929	0.2	2950	1.4	3250	1.0	3600
Industry	27788	1.2	31456	-3.8	28000	1.4	30860	1.3	35120
Building	5730	2.6	7374	-1.3	7100	0.2	7200	0.3	7400
Services	29775	5.0	48298	0.0	48277	2.6	57853	3.1	78362
Total value added (GDP)	66042	3.2	90057	-1.4	86327	2.0	99163	2.3	124482

B. TOTAL FINAL ENERGY DEMAND

2. Luxembourg was a particular case in the preparation of the reference projection because there is as yet no MEDEE demand analysis model for this country. However, sufficient information is available to establish a projection for final energy demand.

Total final energy demand will grow at an annual rate of 1.7% between 1983 and 2000. The GDP growth rate over this period will be 2.2%. The energy intensity of the Luxembourg economy will therefore be declining sharply: from 38.0 (10^3 toe/10^9 LFR (1975)) in 1980 it will fall to 32.7 in 1990 and 28.8 in 2000, making a drop of 24% over 20 years.

Industry will continue to be the main component of final energy demand, with a share of 64% in 2000 against 59% in 1983 and 68% in 1980. Within the sector, the steel industry, which in 1983 accounted for 82% of industrial energy consumption, will continue to dominate the levels and structure of energy consumption throughout the period. A slight increase in output plans up to 1990 will mean an average annual increase in industrial consumption of nearly 4%, mainly solid fuels (coke) and gas (mainly derived gases) and, to a lesser extent, electricity. After 1990, the increase in energy consumption will largely result from the policy for industrial diversification.

Energy consumption in the transport sector is set to rise by 0.8% p.a. between 1983 and 2000, for both passengers and goods. The structure of passenger transport will remain the same throughout the period, both urban and inter-city. Nevertheless, improvements in the specific consumption of vehicles will partly offset the increase in passenger-kilometres between now and 2000.

Energy consumption in the goods transport sector will rise as industrial activity expands, especially in the steel industry. The rise will be greater than in the passenger transport sector and this will increase the share of automotive gas oil in the oil products consumption structure of this sector.

---

[28] AAGR: average annual growth rate.

Energy consumption in the residential and tertiary sector, after levelling off between 1980 and 1983, will resume growth at 1.1% p.a. up to 2000, largely because of rapid expansion in the tertiary sector. The increase in consumption for space heating will be limited, because technical improvements to buildings and heating equipment will to a large extent offset the increase in the volume of premises to be heated. Specific energy consumption, especially electricity, will show a more substantial rise.

Table 2:   Final energy consumption

LUXEMBOURG[29]	Million toe				Annual % changes	
	1980	1983	1990	2000	1990/83	2000/90
Industry[29]	2.32	1.60	2.10	2.30	+4.0	+0.9
Transport	0.50	0.54	0.56	0.62	+0.5	+1.0
Residential and tertiary	0.59	0.56	0.58	0.67	+0.5	+1.4
TOTAL	3.41	2.71	3.24	3.59	+2.6	+1.0
of which						
– solid fuels	1.34	0.88	1.10	1.20	+3.2	+0.9
– oil products	1.08	0.99	1.03	1.13	+0.6	+0.9
– gas	0.69	0.55	0.74	0.79	+4.3	+0.7
– electricity	0.30	0.30	0.37	0.47	+3.0	+2.4

C.

ENERGY SUPPLY

3. Table 3, a breakdown of gross inland primary energy consumption (GIC) based on the economic scenario adopted, shows an increase in GIC of about 2.6% p.a. up to 1990 and 1.0% p.a. thereafter, making an average annual growth rate of 1.6% over the period 1983-2000.

Table 3: Gross inland primary energy consumption

			million toe	
	1983	1990	1995	2000
Coal products	1.27	1.55	1.60	1.65
Oil products	1.00	1.05	1.10	1.15
Natural gas	0.26	0.40	0.42	0.45
Primary electricity	0.30	0.35	0.40	0.45
GIC	2.84	3.35	3.52	3.70

As Luxembourg's energy sector consists of a few small electricity generating units which are integrated into the industrial production structure, GIC is virtually a mirror image of final energy consumption, all requirements being directly covered by imports (including, for coke, oil products and electricity).

---

[29] Including non-energy uses.

D. SUMMARY TABLES

ENERGY SUPPLY / DEMAND BALANCE – 1980   **LUXEMBOURG**

million toe

	SOLID FUEL	PETROL PRODCT	GAS	NUCLR ENERGY	HEAT	HYDRO+ OTHERS	RENEW ENER	TOTAL
PRIM PRODUCT	–	–	–	–	–	0.01	0.01	0.02
TOT IMPORTS	1.84	1.15	0.42	–	–	0.26	–	3.68
TOT EXPORTS	–	0.05	–	–	–	0.02	–	0.07
STOCK CHANGE	0.00	0.00	–	–	–	–	–	0.00
GROSS CONSUM	1.84	1.10	0.42	–	–	0.25	0.01	3.63
BUNKERS	–	–	–	–	–	–	–	–
INLAND CONSM	1.84	1.10	0.42	–	–	0.25	0.01	3.63
ELEC POWER S	0.01	0.02	0.21	–	–	-0.07	0.01	0.18
OTHER TRANSF	0.49	–	-0.49	–	–	–	–	0.0
ENERG INDUST	–	–	0.01	–	–	0.02	–	0.03
FINAL CONSUM	1.34	1.08	0.69	–	–	0.30	–	3.42
– NON ENERGY	–	0.04	–	–	–	–	–	0.04
– INDUSTRY	1.32	0.17	0.59	–	–	0.20	–	2.28
– TRANSPORT	–	0.49	–	–	–	0.01	–	0.50
– HOUSEHOLD	0.02	0.37	0.10	–	–	0.09	–	0.59
STAT DIFFER	–	0.01	–	–	–	–	–	0.01

ENERGY SUPPLY / DEMAND BALANCE - 1983 **LUXEMBOURG**

million toe

	SOLID FUEL	PETROL PRODCT	GAS	NUCLR ENERGY	HEAT	HYDRO+ OTHERS	RENEW ENER	TOTAL
PRIM PRODUCT	-	-	-	-	-	0.01	0.02	0.03
TOT IMPORTS	1.25	1.02	0.26	-	-	0.32	-	2.84
TOT EXPORTS	-	0.02	-	-	-	0.03	-	0.05
STOCK CHANGE	0.02	0.00	-	-	-	-	-	0.02
GROSS CONSUM	1.27	1.00	0.26	-	-	0.30	0.02	2.84
BUNKERS	-	-	-	-	-	0.40	-	-
INLAND CONSM	1.27	1.00	0.26	-	-	0.30	0.02	2.84
ELEC POWER S	0.02	0.01	0.08	-	-	-0.03	0.02	0.10
OTHER TRANSF	0.37	-	-0.37	-	-	-	-	0.00
ENERG INDUST	-	-	-	-	-	0.03	-	0.03
FINAL CONSUM	0.28	0.99	0.55	-	-	0.30	-	2.71
- NON ENERGY	-	0.03	-	-	-	-	-	0.03
- INDUSTRY	0.85	0.11	0.42	-	-	0.19	-	1.57
- TRANSPORT	-	0.53	-	-	-	0.01	-	0.54
- HOUSEHOLD	0.03	0.30	0.13	-	-	0.10	-	0.56
STAT. DIFFER	-	0.01	-	-	-	-	-	0.01

ENERGY SUPPLY / DEMAND BALANCE - 1990

million toe

	SOLID FUEL	PETROL PRODCT	GAS	NUCLR ENERGY	HEAT	HYDRO+ OTHERS	RENEW ENER	TOTAL
PRIM PRODUCT	-	-	-	-	-	0.04	-	0.04
TOT IMPORTS	1.55	1.05	0.40	-	-	0.36	-	3.36
TOT EXPORTS	-	-	-	-	-	0.05	-	0.05
STOCK CHANGE	-	-	-	-	-	-	-	-
GROSS CONSUM	1.55	1.05	0.40	-	-	0.35	-	3.35
BUNKERS	-	-	-	-	-	-	-	-
INLAND CONSM	1.55	1.05	0.40	-	-	0.35	-	3.35
ELEC POWER S	0.10	0.02	0.01	-	-	-0.04	-	0.09
OTHER TRANSF	0.35	-	-0.35	-	-	-	-	0.0
ENERG INDUST	-	-	-	-	-	0.02	-	0.02
FINAL CONSUM	1.10	1.03	0.74	-	-	0.37	-	3.24
- NON ENERGY	-	-	-	-	-	-	-	-
- INDUSTRY	1.05	0.23	0.59	-	-	0.23	-	2.10
- TRANSPORT	-	0.55	-	-	-	0.01	-	0.56
- HOUSEHOLD	0.05	0.25	0.15	-	-	0.13	-	0.58

ENERGY SUPPLY / DEMAND BALANCE - 1995   **LUXEMBOURG**

million toe

	SOLID FUEL	PETROL PRODCT	GAS	NUCLR ENERGY	HEAT	HYDRO+ OTHERS	RENEW ENER	TOTAL
PRIM PRODUCT	-	-	-	-	-	0.05	-	0.05
TOT IMPORTS	1.60	1.10	0.42	-	-	0.40	-	3.52
TOT EXPORTS	-	-	-	-	-	0.05	-	0.05
STOCK CHANGE	-	-	-	-	-	-	-	-
GROSS CONSUM	1.60	1.10	0.42	-	-	0.40	-	3.52
BUNKERS	-	-	-	-	-	-	-	-
INLAND CONSM	1.60	1.10	0.42	-	-	0.40	-	3.52
ELEC POWER S	0.10	0.02	0.01	-	-	-0.04	-	0.09
OTHER TRANSF	0.35	-	-0.35	-	-	-	-	0.00
ENERG INDUST	-	-	-	-	-	0.02	-	0.02
FINAL CONSUM	1.15	1.08	0.76	-	-	0.42	-	3.41
- NON ENERGY	-	-	-	-	-	-	-	-
- INDUSTRY	1.10	0.25	0.60	-	-	0.25	-	2.20
- TRANSPORT	-	0.58	-	-	-	0.01	-	0.59
- HOUSEHOLD	0.05	0.25	0.16	-	-	0.16	-	0.62

ENERGY SUPPLY / DEMAND BALANCE - 2000

million toe

	SOLID FUEL	PETROL PRODCT	GAS	NUCLR ENERGY	HEAT	HYDRO+ OTHERS	RENEW ENER	TOTAL
PRIM PRODUCT	-	-	-	-	-	0.06	-	0.06
TOT IMPORTS	1.65	1.15	0.45	-	-	0.44	-	3.69
TOT EXPORTS	-	-	-	-	-	0.05	-	0.05
STOCK CHANGE	-	-	-	-	-	-	-	-
GROSS CONSUM	1.65	1.15	0.45	-	-	0.45	-	3.70
BUNKERS	-	-	-	-	-	-	-	-
INLAND CONSM	1.65	1.15	0.45	-	-	0.45	-	3.70
ELEC POWER S	0.10	0.02	0.01	-	-	-0.04	-	0.09
OTHER TRANSF	0.35	-	-0.35	-	-	-	-	0.00
ENERG INDUST	-	-	-	-	-	0.02	-	0.02
FINAL CONSUM	1.20	1.13	0.79	-	-	0.47	-	3.59
- NON ENERGY	-	-	-	-	-	-	-	-
- INDUSTRY	1.15	0.27	0.61	-	-	0.27	-	2.30
- TRANSPORT	-	0.61	-	-	-	0.01	-	0.62
- HOUSEHOLD	0.05	0.25	0.18	-	-	0.19	-	0.67

Summarized Energy Balance - LUXEMBOURG

in million toe	1973[a]	1980[a]	1983[a]	1990[b]	1995[b]	2000[b]
I. Gross Energy Consumption	4.50	3.62	2.84	3.35	3.52	3.70
- Bunkers	-	-	-	-	-	-
- Inland consumption	4.50	3.62	2.84	3.35	3.52	3.70
II. Inland Energy Consumption	4.50	3.62	2.84	3.35	3.52	3.70
- Solid fuels	2.45	1.84	1.27	1.55	1.60	1.65
- Oil	1.65	1.10	0.99	1.05	1.10	1.15
- Gas	0.22	0.42	0.26	0.40	0.42	0.45
- Primary electricity, etc.	0.18	0.26	0.32	0.35	0.40	0.45
III.Indigenous Production[1]	0.01	0.02	0.03	0.05	0.07	0.10
- Hard coal	-	-	-	-	-	-
- Lignite & peat	-	-	-	-	-	-
- Oil	-	-	-	-	-	-
- Natural gas	-	-	-	-	-	-
- Nuclear energy	-	-	-	-	-	-
- Hydro & geothermal[2]	0.01	0.01	0.01	0.01	0.01	0.01
- Others & renewables	0.00	0.01	0.02	0.03	0.04	0.05
IV. Net Imports[3]	4.50	3.60	2.79	3.31	3.47	3.64
- Solid fuels	2.46	1.84	1.25	1.55	1.60	1.65
- Oil	1.65	1.10	0.99	1.05	1.10	1.15
- Natural gas[2]	0.22	0.42	0.26	0.40	0.42	0.45
- Electricity[4]	0.17	0.24	0.28	0.31	0.35	0.39
V. Stock Changes[4]	0.01	0.00	-0.02	-	-	-
- Solid fuels	0.01	0.00	-0.02	-	-	-
- Oil	-	0.00	0.00	-	-	-
- Gas	-	-	-	-	-	-
VI. Electricity Generation Input	0.45	0.26	0.13	0.16	0.17	0.18
- Solid fuels[5]	0.29	0.15	0.09	0.10	0.10	0.10
- Oil	0.11	0.02	0.01	0.02	0.02	0.02
- Natural gas	0.04	0.07	0.00	0.01	0.01	0.01
- Nuclear energy	-	-	-	-	-	-
- Hydro & geothermal[2]	0.01	0.01	0.01	0.03	0.04	0.05
- Others & renewables	0.00	0.01	0.02	-	-	-

Main indicators (related to long term objectives)

	1973-1963	1980-1975	1983-1980	1990-1983	2000-1990
Inland energy annual growth rates	+3.5%	-1.2%	-7.7%	+2.6%	+0.9%
GDP annual growth rates	+4.9%	+2.5%	-1.7%	+2.0%	+2.2%
Energy-GDP ratio	0.71	-0.48	0.22	1.30	0.39

	1973	1980	1983	1990	1995	2000
Share of oil in gross energy consumption	36.7%	30.4%	34.8%	30.9%	33.1%	31.1%
Share of coal and nuclear in in electricity production	64.5%	57.7%	69.2%	62.5%	58.8%	55.6%
Supply dependance on imports	100%	99.4%	98.0%	98.8%	98.6%	98.4%

Sources: a. Statistical Office of the European Communities
         b. "Energy 2000"
Notes: 1. Production of primary sources, including recovered products.
         2. The conversion of electricity, including hydro and geothermal, is
            based on its actual energy content: 3600 kjoules/kWh or 860 kcal/kWh
         3. The (-) sign means net exports
         4. The (-) sign means a stock decrease
         5. Including coke oven gas and blast furnace gas (derived from coal)

T H E   N E T H E R L A N D S

A.   ECONOMIC REFERENCE FRAMEWORK

1.   For the period 1983 to 2000, the GDP growth assumption is an
average annual rate of 2.5%. This will be spread unevenly over the
period: 1.9% per annum on average from 1983 to 1990 and 2.8% per
annum from 1990 to 2000.

Between 1983 and 1990, only very small increases in public
consumption (+0.6% per annum) and in public investment (+0.6% per
annum) are expected.   Private consumption will also continue to
stagnate (+0.2% per annum) as disposable incomes fall.
Consequently, GDP growth will be led by gross fixed capital
formation (+2% per annum) and exports (+3.7% per annum). The fall
in interest rates and the livelier world economy will boost firms'
propensity to invest.   Investment in housebuilding fell sharply in
the early 1980s but will improve thereafter (up by 1.5% per annum).
Stimulated by the improvement in world trade, exports are expected
to increase by 3.7% per annum against a 2% per annum rise in
imports.   The traditional exporting industries such as chemicals,
mining and transport should benefit most from the more buoyant
world market.

The trends over the period 1983 to 1990 will continue from 1991 to
2000, but the improvement will be sharper. The GDP growth rate, for
instance, will be close to 3%.   This improvement will be due not
only to the factors operating in the preceding period but also to
the revival of private sector consumption and of demand for
housing.

At the sectoral level, the assumptions adopted point to the
following results:

- consumer goods industries like agri-foodstuffs and textiles will
  remain weak throughout the period, especially from 1983 to 1990,
  owing to the stagnation of disposable income and the consequent
  weakening of the home market;

- the relatively low level of investment in the public sector and
  the depressed demand for housebuilding will affect both building
  and non-metallic minerals;

- the capital goods industry will remain on a sound path throughout
  the period, helped by firms' increased willingness to invest;

- the chemicals industry will also record encouraging results,
  thanks to its export capacities;

- the shift towards a services-based society will continue as
  services expand their share of the economy from 59% in 1983 to
  63% in 2000.

Overall, growth will thus be led mainly by the services, capital
goods and chemicals industries, which accounts for their growing
shares of the value added generated by the productive system
(Table 1).

Table 1: Sectoral trends

HFL '000 million (1975)

	1970	AAGR[30]	1980	AAGR	1983	AAGR	1990	AAGR	2000
Agriculture	7.7	3.9	11.3	6.4	13.6	-1.1	12.5	1.5	14.5
Industry	52.6	3.0	70.7	-1.9	66.7	2.1	77.0	2.2	96.1
EEII[31]	20.7	0.9	22.7	-2.4	21.1	2.8	25.6	2.3	32.2
Other indust.	31.9	4.2	48.0	-1.7	45.6	1.7	51.4	2.2	63.9
Building	13.8	-1.0	12.5	-4.5	10.9	1.6	12.2	2.0	14.8
Services	97.2	3.2	133.8	-0.3	132.7	2.1	153.7	3.3	212.6
Total value added (GDP)	171.3	2.9	228.3	-0.6	223.9	1.9	255.4	2.8	338.0

B.   TOTAL FINAL ENERGY DEMAND

2.   Total final energy demand is expected to grow at an average rate of
     1.5% p.a. between 1983 and 2000, led mainly by consumption in
     industry (1.6% p.a.) and also in the transport sector (1.0% p.a.).
     Energy consumption in the residential and tertiary sector will
     remain relatively stable (0.4% p.a.) and its share of total demand
     will decline from 40% in 1983 to 36% in 2000.

     The relatively high elasticity of industry's energy consumption to
     total industrial value added (0.81 over 1983-2000) can be
     attributed to the absence of any major structural changes in the
     mix of heavy and light industries.

     With passenger traffic strong in the private transport sector and
     inland goods transport still keeping pace with industrial activity,
     there will be no further weakening in the link between energy
     consumption in the transport sector and economic growth.

     The link between energy and economic activity is very weak in the
     residential and tertiary sector, where the nature of the stock of
     housing and office buildings allows substantial energy savings.

     Gas remains the most widely used fuel, mostly for applications in
     the domestic sector. This allows oil products and coal to maintain
     or even increase their shares of energy consumption by industry.
     Generally, the use of electricity increases only with the growth of
     specific requirements.

a)   Industry

3.   The future course of industry will make for increased energy
     consumption, since activity in industries heavily dependent on
     energy is expected to remain relatively dynamic. As a result,

---

[30] AAGR: average annual growth rate.
[31] EEI: energy-intensive industries: metals, non-metallic minerals, paper,
chemicals.

demand for energy will grow by 1.6% p.a. compared with 2.2% annual growth in industrial value added. Despite this close link, the energy intensity[32] of the industrial sector as a whole will fall by 19% between 1980 and 2000, from 197 (10^3 toe/10^9 HFL(1975)) in 1980 to 177 in 1990 and 159 in 2000.

4.   The production plans of the energy-intensive industries envisage continued expansion for all products, especially basic chemicals; as a result these industries will account for much the same share of energy consumption in 2000 as in 1983.

Apart   from   these   industries,   industrial   development   will concentrate on capital goods and chemicals, other than basic chemicals. This too will influence energy consumption, since the "other chemicals" sector has the highest energy intensity of all light industry.

Even if large gains can be expected in furnace end-uses thanks to increased efficiency and higher utilization rates, demand for fuels for furnaces will remain largely unchanged throughout the period: the increase in the tonnages manufactured will counterbalance the significant technological improvements which will allow, for example, a fuel saving of 20% per tonne of steel produced and 25% per tonne of cement.

Steam requirements, especially in the light industries, will continue to grow, though more slowly than the value added of these industries, as more efficient production methods are introduced.

5.   This increase in fuel consumption in industry will be covered, in descending order, by oil products, coal and gas, though gas is reserved primarily for domestic uses.   Electricity will begin to capture heating markets by 2000, but it will continue to be used mainly to meet specific requirements (increasing at an average annual rate of 1.8%).

b)   Transport

6.   Energy consumption will increase significantly for both passenger and goods transport, albeit for different reasons.

Passenger traffic is expected to increase by 2.4% a year in urban zones and by 2.2% on the inter-city routes. It is already very dense, but public transport will carry only 8.5% of the traffic in the cities and only 20% of inter-city traffic in 2000, very close to today's figures.

This situation will boost fuel consumption by private cars: despite a 20% reduction in specific consumption, demand for fuel is expected to grow by 0.8% p.a. on average from 1980 to 2000.

---

[32] Defined here as the ratio of final energy demand (in 10^3 toe) to total value added by industry (in 10^9 HFL at 1975 prices).

Goods traffic will grow by 1.2% p.a., most of it carried by road (80% in 2000). Improvements in the specific consumption of lorries will hold the annual rate of increase in energy consumption to 1.0%.

Overall, passenger traffic's share of total energy consumption in the ransport sector will remain stable over the period. The energy intensity of the sector[33] will fall by about 20% from 1980 to 2000, from 37.7 (10^3 toe/10^7 HFL (1975)) to 35.2 in 1990 and 29.9 in 2000.

c)   Residential and tertiary sector

7.   Given a sustained energy-saving campaign, it will be possible to break the link between energy consumption growth in the residential and tertiary sector and general economic growth. Energy consumption in this sector will grow at an average annual rate of 0.4% from 1983 to 2000, against 2.5% for GDP. The energy intensity[34] of the sector reflects this uncoupling: from 92 (10^3 Toe/10^6 HFL (1975)) in 1980, it will fall to 76 in 1990 and 57 in 2000, a fall of 38% in 20 years.

In 1980 the housing stock was made up mainly of one-family houses, a feature which could even be accentuated in 2000, by which time houses could account for 73% of the stock as against 69% in 1980. Half the houses in 2000 will be "new" houses, built or renovated after 1975. These houses, like the older ones, will be designed and fitted out to allow substantial energy savings, ranging from 15% in "new" houses to 30% in older houses. In older apartment blocks, which will outnumber new ones in 2000, heat savings of about 30% should be made.

These factors suggest that on average home-heating requirements in 2000 will be 20% lower than in 1980, with a corresponding decline in demand for fuel oil in 2000 for this key use in both the residential and tertiary sectors.

Other uses (domestic hot water, specific electricity and cooking) will account for the overall increase in energy consumption in the residential and tertiary sector: specific electricity consumption will grow at an average annual rate of 3%, mainly in response to tertiary sector buoyancy.

Gas will continue to be the most widely used energy vector in this sector up to 2000, leaving a small market share to district heating.

---

[33] Defined here as the ratio of energy consumption in the transport sector to GDP.

[34] Defined here as the ratio of energy consumption in the residential and tertiary sector to GDP.

Table 2:   Final energy consumption

	Million toe				Annual % change	
	1980	1983	1990	2000	1990/83	2000/90
Industry	13.9	11.6	13.6	15.3	+2.3	+1.2
Transport	8.6	8.6	9.0	10.1	+0.7	+1.2
Residential and tertiary	21.0	18.2	19.4	19.4	+0.9	+0.0
Non-energy uses	8.4	6.8	8.6	10.2	+3.4	+1.7
Total	51.9	45.4	50.6	55.0	+1.6	+0.8
- solid fuels	1.4	1.2	1.4	1.4	+2.2	+0.0
- oil products	20.9	17.6	19.5	21.6	+1.5	+1.0
- gas	24.5	21.4	23.6	24.5	+1.4	+0.4
- electricity	5.0	4.9	5.2	6.1	+0.9	+1.6
- other[35]	0.2	0.2	0.9	1.4	+24.0	+4.5

C.   ENERGY SUPPLY

8.   Table 3, a breakdown of gross inland primary energy consumption
(GIC) based on the economic scenario adopted, shows an increase of
some 1.5% per annum from 1983 to 1990 and 0.9% per annum
thereafter, making an average annual growth rate of 1.3% over the
period 1983-2000. Up to 2000, natural gas will remain the largest
component in the energy balance, although its share will fall from
51% in 1983 to 44% in 1990 and 40% in 2000 as output from
indigenous sources declines. Liquid fuels will hold a stable 37%
share. Solid fuels will gain from this change, their share rising
from 10% in 1983 to 16% in 1990 and 15% in 2000. Exhaustion of
natural gas reserves combined with the increase in energy
consumption will bring the degree of dependence on imported energy
up from 7% in 1983 to 26% in 1990 and 42% in 2000.[36] A
source-by-source analysis of these aggregate energy supply trends
reveals movements which are much more marked.

Table 3: Gross inland primary energy consumption

million toe

Form of energy	1983	1990	1995	2000
Coal products	5.9	10.4	11.3	10.5
Oil products	20.5	23.3	24.5	25.9
Natural gas	29.1	28.0	28.9	28.0
Nuclear fuels	0.9	0.9	0.9	4.4
Electricity	0.4	0.3	0.2	-
Renewables	0.2	0.5	0.7	0.8
GIC	57.0	63.4	66.6	69.6

---

[35] Heat and renewable energy sources.
[36] If the present natural gas export contracts are extended up to 2000
(the expected outcome of the 1985 renegotiations), dependence on
energy imports in 2000 will be only 32%.

a)   Oil products

9. While maintaining a steady 37% share of GIC, oil product
    consumption will increase over the period, as the net result of
    several trends:

    - sharply reduced consumption of extra heavy fuel oil in power
      stations, following the conversion to coal of several dual-fired
      (oil/gas) power stations;

    - a continuing low level of heavy fuel oil sales to industry;

    - a rising offtake of medium products as diesel fuel sales
      increase in response to growing total demand in the transport
      sector and the progressive switch to diesel-fuelled motor
      vehicles.

    - increasing sales of petrochemical cuts as this sector expands;

    - declining petrol sales due to the switch to diesel-fuelled motor
      vehicles and the growth of kerosene sales for air transport.

10. Combined with the increase in total consumption of oil products,
    growing sales of medium and light products will add to their share
    of inland consumption, which will rise from 24.8% in 1983 to 37.9%
    in 2000 for medium products and from 40.3% in 1983 to 41% in 2000
    for light products. Since, in the structure of the industry, this
    change will occur primarily at the expense of heavy products, more
    use will have to be made of conversion plant (catalytic cracking,
    hydrocracking and flexicoking) in order to meet inland demand.

Table 4: Structure of inland consumption of oil products
%

Products	1980	1983	1990	1995	2000
Gaseous products	7.9	13.4	10.4	10.3	9.8
Light products	34.6	40.3	40.9	40.2	41.0
Medium products	24.0	24.8	29.9	32.3	37.9
Heavy products	33.5	21.5	18.8	17.2	11.3

As the Dutch refining industry has always been geared to the
exportation of refined products, due account must be taken of this
component in analysing the structure of production in the sector.
Assuming that today's export market trends grow stronger, there
should be a clear shift towards lighter products.

Table 5:   Structure of production in the oil industry (including bunkers)

%

Products	1980	1983	1990	1995	2000
Gaseous products	3.2	5.0	3.6	3.6	3.5
Light products	27.4	35.9	38.2	38.4	39.3
Medium products	33.9	27.0	32.3	32.5	34.2
Heavy products	35.4	32.1	25.9	25.5	23.0

On the basis of the refining capacity in service on 1 January 1984 - 77.6 million tonnes of topping capacity, 8.1 million tonnes of catalytic cracking and thermal cracking, plus 6.0 million tonnes of visbreaking - the Dutch refining industry has a large surplus distillation capacity going back to the first oil shock. Rationalization measures were introduced without delay to bring refining capacity down from 102 million tonnes in 1975 to 85.6 million tonnes in 1978 and 77.6 million tonnes in 1983. But, at the same time, crude oil inputs into refineries slumped from 57 million toe in 1975 and 1978 to 43.7 million toe in 1983, thus holding the capacity utilization rate at 56%.

Unless large new export markets can be found, which seems far from likely in view of fresh competition from the oil-producing countries, the current rationalization programme will have to be stepped up in order to close 15-20 million tonnes of capacity by 1990.

b)   Natural gas

11.   Natural gas is the Netherlands' leading energy resource, accounting for over 51% of GIC in 1983. However, the next 20 years will bring a steady reduction in the volumes produced from the national fields, from 66.7 million toe in 1980 to 49.7 million toe in 1990 and 39.1 million toe in 2000. The volumes exported will be steadily reduced.[37]   It is also assumed that imports from Norway will increase only slightly up to the end of the century, from 2.4 million toe in 1983 to 4 million toe in 1995.

12.   The structure of inland natural gas consumption will reflect current trends: a reduced offtake by power stations, lower sales to the industrial sector and no change in sales to the residential and tertiary sector, where the market is saturated and there are only limited prospects of any fresh penetration.

In contrast to the situation in countries which do not produce natural gas, significant quantities will be delivered to power stations right through until 1990. There are two reasons for this: first, the downward revision of the energy consumption forecasts for all European countries will restrict export opportunities for Dutch gas, which will have to find new outlets on the home market; secondly, the sale of large quantities of natural gas to the

---

[37] Not allowing for the 1985 renegotiations.

electricity industry will reduce electricity's dependence on oil and enable the industry to anticipate the impact of conversion to coal of a significant slice of generating capacity.

c)   Coal products

13. Coal products are attractively enough priced for consumption to rise strongly in the medium and long term, (+4.5 million toe). Their share of inland energy consumption will thus increase from 10% in 1983 to 16% in 1990 and 15% in 2000. Since, in addition, national resources at great depths cannot be worked economically, this increase will have to be covered by imports alone, which in turn will require infrastructure improvements for harbours, storage facilities and so on, and an expanded inland transport network.

This increase in consumption is of course due to the switch to coal by the electricity industry. Some 9 million toe of coal will be burned for electricity generation in 1995, against 2.9 million toe in 1983.

In addition to the role played by the electricity sector, steam coal consumption in industry should rise in the long term, particularly in cement works.

d)   Electricity sector

14. The basic feature of the electricity sector is coal's increasing share of demand, which will grow at an annual rate of 1.1% from 1983 to 1990 and 1.6% from 1991 to 2000, making an average annual growth rate of 1.4% over the whole period.

Before discussing the investments to be made in electricity generating capacity, it is important to specify the assumptions adopted for nuclear power stations. All investment in nuclear power stations has been frozen until the public debate on energy is closed. Without wishing to anticipate the outcome of the debate, we adopted one of the alternatives put forward, namely the development of 2600 MW of nuclear power by 2000. In addition to these investments, a major power station conversion effort is in hand, from dual-fired (oil/gas) to coal: 1600 MW between 1981 and 1990, 310 MW between 1991 and 1995 and 300 MW between 1996 and 2000; and the expansion of CHP generation is already an important feature of the electricity sector's investment programme.

Table 6: Capital investment programme for the electricity sector

Unit	1981-1990	1991-1995	1996-2000
Nuclear	-	-	2600 Mwe
Thermal	2320 Mwe	255 Mwe	720 Mwe
Wind power	31 Mwe	76 Mwe	252 Mwe
CHP	1450 Mwth	1080 Mwth	1225 Mwth

15.   The overall breakdown of electricity production (all producers) by
type of fuel shows the increasing share to be taken by coal: from
1990 onwards, more than 50% of the Netherlands' electricity will be
generated from coal.   In the long run,   nuclear power should also
take its share up to nearly 22% by 2000 (Table 7).   In the short
term liquid fuels will feel the full impact of the competition from
solid fuels,   while in the longer term natural gas will yield to
coal and nuclear.

Table 7:   Total net electricity production

| Origin | 1983 | | 1990 | | 1995 | | 2000 | |
	TWh	%	TWh	%	TWh	%	TWh	%
Nuclear	3.4	5.9	3.3	5.2	3.3	4.6	17.0	22.0
Solid fuels	13.0	22.9	34.4	54.8	39.0	55.9	37.5	48.3
Liquid fuels	5.3	9.4	0.4	0.6	0.5	0.7	0.0	0.0
Gaseous fuels	34.2	60.0	24.4	38.9	26.4	37.9	21.7	28.0
Renewables	1.0	1.8	0.3	0.5	0.6	0.9	1.4	1.7
Total	57.0	100.0	62.8	100.0	69.8	100.0	77.6	100.0

D. <u>SUMMARY TABLES</u>

ENERGY SUPPLY/DEMAND BALANCE - 1980   **NETHERLANDS**

million toe

	SOLID FUEL	PETROL PRODCT	GAS	NUCLR ENERGY	HEAT	HYDRO+ OTHERS	RENEW ENER	TOTAL
PRIM PRODUCT	–	1.6	66.7	1.1	–	–	0.3	69.6
TOT IMPORTS	4.9	79.0	2.9	–	–	0.3	–	87.1
TOT EXPORTS	0.8	41.4	39.1	–	–	0.4		81.7
STOCK CHANGE	–	-0.7	–	–	–	–	–	-0.7
GROSS CONSUM	4.2	38.4	30.4	1.1	–	0.3	–	74.3
BUNKERS	–	9.3	–	–	–	–	–	9.3
INLAND CONSM	4.2	29.1	30.4	1.1	–	0.3	–	65.1
ELEC POWER S	1.4	5.2	5.9	1.1	-0.2	-5.6	0.3	8.1
OTHER TRANSF	1.2	0.4	-1.1	–	–	–	–	0.5
ENERG INDUST	–	2.7	0.8	–	–	0.6	–	4.2
FINAL CONSUM	1.4	20.9	24.5	–	0.2	5.0	–	51.9
– NON ENERGY	0.1	6.6	1.7	–	–	–	–	8.4
– INDUSTRY	1.0	3.0	7.2	–	0.2	2.5	–	13.9
– TRANSPORT	–	8.5	–	–	–	0.1	–	8.6
– HOUSEHOLD	0.3	2.7	15.5	–	–	2.4	–	21.0
STAT DIFFER	0.2	-0.1	0.4	–	–	–	–	0.5

ENERGY SUPPLY / DEMAND BALANCE - 1983

million toe

	SOLID FUEL	PETROL PRODCT	GAS	NUCLR ENERGY	HEAT	HYDRO+ OTHERS	RENEW ENER	TOTAL
PRIM PRODUCT	–	2.9	55.3	0.9	–	–	0.2	59.4
TOT IMPORTS	5.5	77.3	2.4	–	–	0.5	–	85.8
TOT EXPORTS	1.0	51.4	28.6	–	–	0.1	–	81.2
STOCK CHANGE	0.6	1.1	–	–	–	–	–	1.7
GROSS CONSUM	5.1	29.9	29.2	0.9	–	0.4	0.2	65.7
BUNKERS	–	8.0	–	–	–	–	–	8.0
INLAND CONSM	5.1	21.9	29.2	0.9	–	0.4	0.2	57.7
ELEC POWER S	3.0	1.1	7.7	0.9	-0.2	-5.1	0.2	7.6
OTHER TRANSF	0.9	0.4	-0.9	–	–	–	–	0.3
ENERG INDUST	–	2.8	1.0	–	–	0.6	–	4.4
FINAL CONSUM	1.2	17.6	21.4	–	0.2	4.9	–	45.4
– NON ENERGY	0.1	5.4	1.3	–	–	–	–	6.8
– INDUSTRY	1.0	1.5	6.6	–	0.2	2.3	–	11.6
– TRANSPORT	–	8.5	–	–	–	0.1	–	8.6
– HOUSEHOLD	0.1	2.2	13.5	–	–	2.5	–	18.2
STAT DIFFER	–	0.0	0.1	–	–	–	–	0.2

ENERGY SUPPLY / DEMAND BALANCE - 1990

NETHERLANDS

million toe

	SOLID FUEL	PETROL PRODCT	GAS	NUCLR ENERGY	HEAT	HYDRO+ OTHERS	RENEW ENER	TOTAL
PRIM PRODUCT	–	3.7	49.7	0.9	–	–	0.5	54.8
TOT IMPORTS	11.2	48.3	4.0	–	–	0.3	–	63.8
TOT EXPORTS	0.8	17.7	25.7	–	–	–	–	44.2
STOCK CHANGE	–	–	–	–	–	–	–	–
GROSS CONSUM	10.4	34.3	28.0	0.9	–	0.3	0.5	74.4
BUNKERS	–	11.0	–	–	–	–	–	11.0
INLAND CONSM	10.4	23.3	28.0	0.9	–	0.3	0.5	63.4
ELEC POWER S	7.9	0.1	5.5	0.9	-0.8	-5.5	0.4	8.5
OTHER TRANSF	1.1	0.4	-1.5	–	–	–	–	–
ENERG INDUST	–	3.3	0.4	–	0.0	0.6	–	4.3
FINAL CONSUM	1.4	19.5	23.6	–	0.8	5.2	0.1	50.6
– NON ENERGY	–	6.5	2.1	–	–	–	–	8.6
– INDUSTRY	1.3	2.1	7.3	–	0.3	2.6	–	13.6
– TRANSPORT	–	8.9	–	–	–	0.1	–	9.0
– HOUSEHOLD	0.1	2.0	14.2	–	0.5	2.5	0.1	19.4

ENERGY SUPPLY / DEMAND BALANCE - 1995

million toe

	SOLID FUEL	PETROL PRODCT	GAS	NUCLR ENERGY	HEAT	HYDRO+ OTHERS	RENEW ENER	TOTAL
PRIM PRODUCT	–	3.0	42.9	0.9	–	–	0.7	47.5
TOT IMPORTS	12.1	50.8	4.0	–	–	0.2	–	67.2
TOT EXPORTS	0.8	17.8	18.0	–	–	–	–	36.6
STOCK CHANGE	–	–	–	–	–	–	–	–
GROSS CONSUM	11.3	36.0	28.9	0.9	–	0.2	0.7	78.1
BUNKERS	–	11.5	–	–	–	–	–	11.5
INLAND CONSM	11.3	24.5	28.9	0.9	–	0.2	0.7	66.6
ELEC POWER S	8.8	0.1	5.8	0.9	-1.2	-6.0	0.6	9.0
OTHER TRANSF	1.0	0.4	-1.5	–	–	–	–	–
ENERG INDUST	–	3.7	0.4	–	0.1	0.6	–	4.8
FINAL CONSUM	1.5	20.3	24.2	–	1.1	5.6	0.1	52.8
– NON ENERGY	–	7.0	2.4	–	–	–	–	9.4
– INDUSTRY	1.4	2.0	7.8	–	0.4	2.8	–	14.4
– TRANSPORT	–	9.4	–	–	–	0.1	–	9.5
– HOUSEHOLD	0.1	1.9	14.0	–	0.7	2.7	0.1	19.5

ENERGY SUPPLY / DEMAND BALANCE   –   NETHERLANDS
                                       million toe

	SOLID FUEL	PETROL PRODCT	GAS	NUCLR ENERGY	HEAT	HYDRO+ OTHERS	RENEW ENER	TOTAL
PRIM PRODUCT	–	3.0	39.1	4.4	–	–	0.8	47.3
TOT IMPORTS	11.3	52.9	2.6	–	–	–	–	66.8
TOT EXPORTS	0.8	18.0	13.7	–	–	–	–	32.5
STOCK CHANGE	–	–	–	–	–	–	–	–
GROSS CONSUM	10.5	37.9	28.0	4.4	–	–	0.8	81.6
BUNKERS	–	12.0	–	–	–	–	–	12.0
INLAND CONSM	10.5	25.9	28.0	4.4	–	–	0.8	69.6
ELEC POWER S	8.2	0.0	4.6	4.4	-1.4	-6.8	0.7	9.7
OTHER TRANSF	0.9	0.5	-1.4	–	–	–	–	–
ENERG INDUST	–	3.8	0.3	–	0.1	0.7	–	4.9
FINAL CONSUM	1.4	21.6	24.5	–	1.3	6.1	·0.1	55.0
– NON ENERGY	–	7.8	2.4	–	–	–	–	10.2
– INDUSTRY	1.4	2.1	8.2	–	0.5	3.1	–	15.3
– TRANSPORT	–	10.0	–	–	–	0.1	–	10.1
– HOUSEHOLD	–	1.7	13.9	–	0.8	2.9	0.1	19.4

Summarized Energy Balance - NETHERLANDS

in million toe	1973[a]	1980[a]	1983[a]	1990[b]	1995[b]	2000[b]
I. Gross Energy Consumption	72.89	74.34	65.74	74.40	78.10	81.60
- Bunkers	11.53	9.26	8.05	11.00	11.50	12.00
- Inland consumption	61.36	65.08	57.69	63.40	66.60	69.60
II. Inland Energy Consumption	61.36	65.08	57.69	63.40	66.60	69.60
- Solid fuels	3.16	4.16	5.13	10.40	11.30	10.50
- Oil	29.51	29.15	21.86	23.30	24.50	25.90
- Gas	28.50	30.42	29.17	28.00	28.90	28.00
- Primary electricity, etc.	0.19	1.36	1.53	1.70	1.80	5.20
III. Indigenous Production[1]	56.78	69.64	59.40	54.80	47.50	47.30
- Hard coal	1.19	-	-	-	-	-
- Lignite & peat	-	-	-	-	-	-
- Oil	1.54	1.58	2.93	3.70	3.00	3.00
- Natural gas	53.75	66.67	55.34	49.70	42.90	39.10
- Nuclear energy	0.30	1.07	0.90	0.90	0.90	4.40
- Hydro & geothermal[2]	-	-	-	-	-	-
- Others & renewables	-	0.32	0.23	0.50	0.70	0.90
IV. Net Imports[3]	16.28	5.36	4.61	19.60	30.60	34.30
- Solid fuels	1.69	4.12	4.47	10.40	11.30	10.50
- Oil	39.96	37.52	25.91	30.60	33.00	34.90
- Natural gas[2]	-25.25	-36.25	-26.17	-21.70	-14.00	-11.10
- Electricity[2]	-0.12	-0.03	0.40	0.30	0.20	-
V. Stock Changes[4]	0.18	0.65	-1.73	-	-	-
- Solid fuels	-0.28	-0.04	-0.66	-	-	-
- Oil	+0.46	0.69	-1.07	-	-	-
- Gas	-	-	-	-	-	-
VI. Electricity Generation Input	12.03	13.93	12.88	16.80	18.20	19.70
- Solid fuels[5]	0.81	1.71	3.31	8.90	9.80	9.10
- Oil	1.55	5.24	1.10	0.10	0.10	0.00
- Natural gas	9.37	5.59	7.34	6.50	6.80	5.50
- Nuclear energy	0.30	1.07	0.90	0.90	0.90	4.40
- Hydro & geothermal[2]	-	-	-	-	-	-
- Others & renewables	-	0.32	0.23	0.40	0.60	0.70

Main indicators (related to long term objectives)

	1973-1963	1980-1975	1983-1980	1990-1983	2000-1990
Inland energy annual growth rates	+8.3%	+2.1%	-3.9%	+1.3%	+0.9%
GDP annual growth rates	+5.5%	+2.5%	-0.6%	+2.0%	+2.8%
Energy-GDP ratio	1.51	0.84	-	0.67	0.34

	1973	1980	1983	1990	1995	2000
Share of oil in gross energy consumption	56.3%	51.7%	45.5%	46.1%	46.1%	46.4%
Share of coal and nuclear in in electricity production	9.2%	20.0%	32.7%	58.3%	58.8%	68.5%
Supply dependance on imports	22.3%	7.2%	7.0%	26.3%	39.2%	42.0%

Sources: a. Statistical Office of the European Communities
        b. "Energy 2000"

Notes: 1. Production of primary sources, including recovered products.
      2. The conversion of electricity, including hydro and geothermal, is based on its actual energy content: 3600 kjoules/kWh or 860 kcal/kWh
      3. The (-) sign means net exports
      4. The (-) sign means a stock decrease
      5. Including coke oven gas and blast furnace gas (derived from coal)

U N I T E D   K I N G D O M

A. ECONOMIC REFERENCE FRAMEWORK

1. Gross domestic product started to recover in the summer of 1982 and
    is assumed to continue to grow to 1990, initially as both household
    consumption and housing investment increase and then under the
    stimulus of rising productive investment as company trading results
    improve. In spite of somewhat gloomier prospects for internal
    demand generally in the late 1980s, the vigorous growth of GFCF
    will underpin economic recovery throughout the second half of the
    decade. GDP is expected to grow at an average annual rate of 2.2%
    over the period 1983-90.

    These trends will broadly continue over the period 1991-2000,
    boosting the average annual growth rate of GDP to 2.5%.

    Investment will grow in all sectors including market services, but
    unevenly across the whole range of industry. This will make for a
    general restructuring of the productive system, but certain heavy
    and consumer goods industries seem likely to remain very
    uncompetitive.

    Expanding investment will be a feature of the capital goods and
    building industries and boost activity in these sectors: their
    rate of growth will be faster than the average of the others. For
    reasons more to do with the availability of natural resources, the
    basic chemicals industry will continue its fairly vigorous
    expansion. Service sector activity will of course be stimulated by
    these sectoral developments.

    In addition, a more favourable external trade trend in the medium
    and long term should aid economic recovery after 1985. Although
    imports will continue to increase at a relatively high rate,
    exports should benefit from a more propitious international
    economic environment. Capital goods and market services will lead
    the way as energy resources become scarcer. Table 1 shows these
    sectoral movements.

Table 1: Sectoral trends

UKL million (1975)

	1970	AAGR[38]	1980	AAGR	1983	AAGR	1990	AAGR	2000
Agriculture	1950	2.1	2400	2.3	2573	3.0	3171	0.0	3171
Industry	23393	2.7	30433	0.5	30930	3.5	39326	1.5	45583
EII[39]	5000	4.6	7854	0.2	7909	2.3	9257	1.9	11182
Other indust.	18393	2.1	22579	0.6	23021	3.9	30069	1.4	34401
Building	6990	-0.2	6818	-1.6	6498	2.5	7728	3.0	10386
Services	57200	1.9	69016	2.0	73143	1.5	81070	3.0	108992
Total value added (GDP)	89533	2.0	108667	1.4	113144	2.1	131295	2.5	168132

---

[38] AAGR: average annual growth rate.
[39] energy-intensive industries: metals, non-metallic minerals,   paper,
    chemicals.

B.   TOTAL FINAL ENERGY DEMAND

2.   Total final energy demand will exhibit a relatively slow growth rate between 1983 and 2000 (annual average of 1.1%) compared with the economic growth rate, which will average 2.3% per annum.

This significant decoupling of energy demand from economic growth will be the result of:

- structural changes in the housing stock and industrial production. The structure of industry will shift towards light industries and a larger proportion of housing will be apartment blocks; and

- technical improvements in plant and equipment and better insulation of buildings, which will lead to unit energy consumption savings (per housing unit or per tonne of goods produced) of 10%-15% in 2000.

3.   Consumption of gas and electricity will increase the fastest (1.2% and 1.8% per year respectively between 1983 and 2000). The increase in the use of gas will be similar in industry and in the residential sector. The rise in electricity consumption will be primarily due to the increase in specific electricity requirements.

The demand for oil products will rise slightly, as a result of increased consumption in the transport sector.

(a)   Industry

4.   Despite the firm recovery of industry, with an average annual growth rate of 2.3% from 1983 to 2000, energy consumption in this sector will rise only slightly, at an average annual rate of around 1.4%. This trend will make for a marked and continuous fall in the total energy intensity[40] of the industrial sector, from 1.19 toe/1000 UKL in 1980 to 1.10 in 1990 and 1.03 in 2000.

Two factors will contribute to this: changes in economic structure, and technological change.

- Light industries (capital and consumer goods industries) will increase their share in total industrial value added at the expense of the intermediate goods industries (steel, basic chemicals, non-metallic minerals, etc). Since the energy intensity of the former is considerably lower than that of the latter (by a factor of as much as 6 to 10), the effects on energy consumption will be appreciable.

- More energy-thrifty processes will continue to be introduced and technical performances will improve. The majority of cement factories, for example, will switch to the dry method, and more recycled products will be used in the manufacture of paper (15% more in the year 2000). The consumption of steam, heat or

---

[40] Industrial energy intensity is the ratio of energy consumption in industry to industrial added value.

electricity per unit of manufactured product (or unit of value added for light industries) will improve appreciably by 2000: a fuel saving of 13% per tonne of steel produced, and a reduction of 10-15% in the quantity of steam consumed per unit of value added in the textile and agri-foodstuffs industries.

5. The breakdown by fuel of industrial energy consumption will shift in emphasis from oil products to gas, electricity and solid fuels which more or less equally meet the increase in the total energy demand of industry.

The expanding basic chemicals industry, thriving on the abundance of natural resources, will increase non-energy uses at an average annual rate of 1.5% from now until 2000. The increase will solely affect oil products.

(b) Transport

6. Passenger transport demand is closely tied to household incomes. Most inter-city passenger traffic is by car (92% in 1980); and 66% of urban passenger transport is by car. The relative positions are expected to be the same in the year 2000, although private cars will then account for only 88% of inter-city passenger traffic and 54% of urban passenger traffic.

A notable improvement in the specific consumption of cars (about 20%) added to a reduction in the average annual distance travelled (about 9%) will hold down energy consumption by private cars, despite the larger number of cars by the year 2000 (approximately 25 million as against 16 million in 1980). Consumption in this category will rise at an average rate of 0.4% over the period 1980-2000.

7. Goods transport depends primarily on industrial activity. Healthy growth in the productive sector will thus make for a higher growth rate of energy consumption in goods transport (annual average of 1.8%) than in passenger transport. Goods traffic will continue to be carried mainly by road, which will retain the 89% share it held in 1980.

Finally, total consumption in the transport sector, which currently divides 60-40 between passenger transport and goods transport, will shift towards goods transport by the year 2000 (52% - 48%).

(c) Residential and tertiary sector

8. The decoupling of energy consumption from economic growth is expected to be most marked in the residential and tertiary sector: energy consumption will increase only very slightly over the entire period (average 0.4% per annum), while both GDP and household consumption will grow at a rate of 2.3-2.5% p.a. These contrasting trends will make for appreciable gains in the sector's energy intensity,[41] from 0.49 in 1983 to 0.43 in 1990 and 0.35 in 2000, a

---

[41] Defined here as the ratio of energy consumption in million toe to GDP

29% improvement by the end of the period.

The decoupling of the rates for the sector as a whole is confirmed when the residential and tertiary sectors are considered separately. Their respective shares in the total energy consumption of the sector will remain much the same from 1980 to 2000 at 68% and 27% (the remaining 5% being consumption in agriculture).

9.   The structural changes in the housing stock combined with much more efficient insulation in buildings and improved heating installations will make for substantial reductions in heating requirements. For one thing, the proportion of appartment blocks (which consume less energy on average per unit than single-family houses (ratio of 2 to 3)) will be higher in 2000 than in 1980; for another, the insulation of older buildings and stricter standards for new buildings will reduce unit requirements by 8-15% depending on the age and the type of construction. Overall, average heating requirements per dwelling in 2000 will be 10% less than in 1980.

The average requirements per dwelling for other uses will increase slightly for hot water (5% more in 2000) and substantially for specific electricity (35% more in 2000).

In the tertiary sector the improved technical performance of heating equipment and buildings will hold energy consumption stable between now and 2000 despite the substantial economic growth in this sector. The structure of energy uses will change: heating requirements will decline and specific electricity requirements will increase.

10.   Gas will be used more widely for heating and domestic hot water, unlike other fuels, and its share of consumption in the sector will rise from 46% in 1983 to 52% in 2000. This increase in market share will apply both to new buildings and to older ones, of which 15% of those still standing in 2000 will have been converted.

Electricity demand in the residential and tertiary sector will increase at an average annual rate of 1.4% from 1983 to 2000.

---

in 10^9 UKL at 1975 prices.

Table 2: Final energy consumption

| | million toe | | | Annual % change | |
UK	1983	1990	2000	1990/83	2000/90
Industry	36.9	43.3	47.0	+2.3	+0.8
Transport	32.7	36.1	39.5	+1.4	+0.9
Residential & tert	55.2	57.1	58.7	+0.5	+0.3
Non-energy uses	8.4	9.8	12.8	+2.2	+2.7
Total	133.2	146.3	158.0	+1.5	+0.8
of which:					
- Solid fuels	15.4	13.8	13.5	-1.6	-0.2
- Oil products	58.7	63.9	67.2	+1.2	+0.5
- Gas	39.6	45.1	48.8	+1.9	+0.8
- Electricity	19.4	22.3	26.4	+2.0	+1.7
- Others[42]	0.0	1.2	2.1	..	+5.8

C. ENERGY SUPPLY

11. A breakdown of gross inland primary energy consumption (GIC) shows an increase of about 1.4% per annum between now and 1990 and 0.9% per annum thereafter, making an average annual growth rate of 1.1% over the period 1983-2000 (Table 3). The main structural feature will be the growing share of nuclear fuels, from 7% in 1983 to 9.6% in 1990 and 14.8% in 2000, at the expense of solid fuels, whose share will slip from 33.7% in 1983 to 29.9% in 2000. Other fuels will retain more or less the same market shares. The development of hydrocarbon resources and nuclear capacity will enable the United Kingdom to maintain more than 100% energy supply self-sufficiency up to 1990 and still almost 98% in 2000, despite the appreciable increase in energy consumption by then that is assumed.

Table 3: Gross inland primary energy consumption

| | | | million toe | |
Form of energy	1983	1990	1995	2000
Coal products	65.3	72.0	73.4	70.4
Oil products	72.4	74.2	76.9	78.8
Natural gas	42.4	47.1	50.7	50.7
Nuclear fuels	13.5	20.5	22.5	34.8
Hydro	0.4	0.4	0.5	0.5
Renewable energy sources	-	-	-	0.2
GIC	194.0	214.2	224.0	235.4

(a) Oil products

12. Despite a declining share of GIC, oil product consumption will increase slightly throughout the period, as the net result of several trends:

---

[42]Heat and renewable sources of energy.

- lower consumption of extra heavy fuel oil in power stations, where it will face competition from both nuclear power and coal;

- lower sales of heavy fuel oil to industry and the tertiary sector, under the impact of competition on its traditional markets from natural gas, and from coal, which will continue to make inroads in the industrial sectors;

- a slight fall in the use of heating oil by the residential sector and industry, both of which will show a preference for natural gas;

- increasing sales of diesel fuel as the goods transport market grows;

- substantially rising naphtha sales as the petrochemical industry expands;

- growing sales of kerosene for air transport.

13. As the overall consumption of oil products rises, the drop in heavy product sales will reduce their share in the structure of consumption from 33.5% in 1980 to 25.6% in 1990 and 24.2% in 2000 (Table 4). The shares of light products (3.7%) and medium products (5.4%) will increase over the same period. This change in the consumption structure of oil products, which in the United Kingdom is virtually the same as the production structure shows quite clearly the need for a substantial conversion capacity.

Table 4: Structure of inland consumption of oil products

| | | | | | % |
Products	1980	1983	1990	1995	2000
Gaseous products	2.0	3.9	2.4	2.4	2.2
Light products	40.3	44.6	43.4	43.5	44.0
Medium products	24.2	24.9	28.6	28.9	29.6
Heavy products	33.5	26.6	25.6	25.2	24.2

On the basis of refining capacity at 1 January 1983 (113.0 million tonnes of atmospheric distillation, 14.70 million tonnes of catalytic cracking and hydrocracking plus 5.1 million tonnes of thermal cracking, 3.0 million tonnes of coking and 2.0 million tonnes of visbreaking) the United Kingdom has ample conversion capacity to cope with the changes in the consumption structure in the short and medium term, provided that it does not pursue a policy of exporting refined products with the aim of increasing the return on its indigenous crude. On the other hand, the UK already has a surplus of distillation column capacity compared to the quantities of crude to be treated. Restructuring of these units, started in 1980, eliminated 17 million tonnes of capacity in two years. The process will have to continue if a sufficient rate of return is to be obtained across the entire production network after restructuring.

(b)   Natural gas

14.   Up to 1995, the general trend will be a steady increase in the
       inland consumption of natural gas, from 42.4 million toe in 1983
       to 47.1 milliion toe in 1990, 50.7 million toe in 1995 and
       50.7 million toe in 2000. Consequently, the share of gas in GIC
       will remain stable at around 22%. This trend results from maximum
       exploitation of national resources with limited imports of
       Norwegian gas, which in the medium term will mean the purchase of
       gas from the jointly exploited Frigg field. After 1990, under the
       reference scenario, imports must make up the quantities that the
       UK would have obtained from, for example, the Sleipner field.[43]

       In this context, the inland consumption of natural gas will follow
       the pattern of current trends: a steady increase in sales to the
       residential and tertiary sector (29.0 million toe in 1990 and
       30.6 million toe in 2000), consolidation of the market share won
       in the industrial sectors (16.1 million toe in 1990 and
       17.2 million toe in 2000) and, less predictably, expansion in the
       medium term of natural gas deliveries to the electricity sector,
       where natural gas can easily cope with peak power demand and is
       the most suitable fuel for small CHP plants.

(c)   Coal products

15.   Consumption of coal products, national resources of which will
       continue to be extensively exploited, will be rising in 1995
       (72.0 million toe in 1990 and 73.4 million toe in 2000), their
       share in GIC thus remaining close to 33% over the period.

       This rise is explained by the increase in total electricity
       demand, which is expected to grow by 1.9% per annum between 1983
       and 1995 after relative stability between 1975 and 1983
       (+0.1% per annum). This increase in demand will result in
       extensive use of coal-fired power stations to cover the base load
       and consequently an increase of coal deliveries of 9 million toe
       by 1990.

       After 1995 the expanding nuclear programme will reduce sales of
       solid fuels to the electricity sector by 3.5 million toe while
       total consumption falls by 3.0 million toe, bringing the share of
       solid fuels in GIC down to 29.9%.

(d)   Electricity sector

16.   The main feature in the electricity sector will be the increasing
       share of nuclear power in meeting demand, whose annual growth rate
       will be +2.0% between 1983 and 1990, +1.6% between 1991 and 1995,
       and +1.7% between 1996 and 2000, making an average annual growth
       rate of 1.8% over the whole period.

---

[43] Since this study was made, the UK has revised upwards its estimates of
      natural gas reserves and its production plans and has abandoned for
      the moment any purchase of gas from Sleipner.

Investments under the nuclear programme assumed for this study are for the installation by the year 2000 of 19.7 GW. This will not be sufficient both to cover capacity expansion and to replace conventional power stations decommissioned after 35 years of operation. Consequently, the capital investment programme for the electricity sector will also include single-fuel coal-fired power stations (2.4 GW between 1996 and 2000). CHP units burning natural gas and coal and supplying heat to district heating systems will also increase.

Table 5: Capital investment programme for the electricity sector

	1981-90	1991-1995	1996-2000
Nuclear GCR	6.200 MWe	—	—
Nuclear LWR	—	1.200 MWe	7.200 MWe
Coal	1.410 MWe	—	2.420 MWe
Dual-fuel plants	9.015 MWe	—	—
Pumping	1.675 MWe	—	—
CHP	2.500 MWth	2.852 MWth	2.215 MWth

17.   The overall production structure (all producers) by type of fuel shows the large share gained by nuclear power: by the year 2000 some 32% of the electricity produced will be of nuclear origin, while coal-fired production will stabilize at around 210-220 TWh. The share of liquid and gaseous fuels, with no more than 9.4% of production in 1983, will drop to 6.4% in 1990 and 6.1% in 2000. The stabilizing of their share in the medium term is due to increased deliveries of oil and natural gas to power stations, pending the entry into operation of the new nuclear power stations.

Table 6:   Total electricity production

								TWh and %
Origin	1983		1990		1995		2000	
	TWh	%	TWh	%	TWh	%	TWh	%
Hydro	6.4	2.5	4.2	1.4	4.2	1.3	4.2	1.2
Nuclear power	43.9	17.0	60.8	20.7	68.1	21.2	114.2	32.4
Solid fuels	183.6	71.1	210.6	71.5	220.9	68.7	209.4	59.4
Liquid fuels	22.5	8.7	11.8	4.0	15.9	4.9	17.8	5.0
Gaseous fuels	1.8	0.7	7.1	2.4	12.0	3.8	3.8	1.1
Other	0.0	0.0	0.0	0.0	0.3	0.1	3.3	0.9
Total	258.2	100.0	294.5	100.0	321.4	100.0	352.7	100.0

D. SUMMARY TABLES

ENERGY SUPPLY / DEMAND BALANCE - 1980   **UNITED KINGDOM**

million toe

	SOLID FUEL	PETROL PRODCT	GAS	NUCLR ENERGY	HEAT	HYDRO+ OTHERS	RENEW ENER	TOTAL
PRIM PRODUCT	74.7	79.7	30.9	10.4	-	0.3	-	196.1
TOT IMPORTS	5.1	56.2	9.0	-	-	-	-	70.2
TOT EXPORTS	3.3	54.2	-	-	-	-	-	57.5
STOCK CHANGE	-6.6	0.1	-	-	-	-	-	-6.5
GROSS CONSUM	70.0	81.8	39.9	10.4	-	0.3	-	202.3
BUNKERS	-	2.4	-	-	-	-	-	2.4
INLAND CONSM	70.0	79.4	39.9	10.4	-	0.3	-	199.9
ELEC POWER S	50.1	8.1	0.8	10.4	-	-24.2	-	45.3
OTHER TRANSF	2.9	1.1	-2.6	-	-	-	-	1.4
ENERG INDUST	0.4	6.5	2.4	-	-	4.4	-	13.7
FINAL CONSUM	16.0	63.2	38.5	-	-	20.2	-	138.0
- NON ENERGY	0.6	6.8	-	-	-	-	-	7.4
- INDUSTRY	5.9	13.4	14.7 *	-	-	7.4	-	41.4
- TRANSPORT	-	32.7	-	-	-	0.4	-	33.1
- HOUSEHOLD	9.5	10.4	23.8	-	-	12.4	-	56.1
STAT DIFFER	0.3	0.5	0.8	-	-	-	-	1.6

ENERGY SUPPLY / DEMAND BALANCE - 1983

million toe

	SOLID FUEL	PETROL PRODCT	GAS	NUCLR ENERGY	HEAT	HYDRO+ OTHERS	RENEW ENER	TOTAL
PRIM PRODUCT	69.2	117.0	32.8	13.5	-	0.4	-	232.9
TOT IMPORTS	3.5	40.5	9.6	-	-	-	-	53.6
TOT EXPORTS	4.3	84.2	-	-	-	-	-	88.5
STOCK CHANGE	-3.7 (−3.1)	1.0	-	-	-	-	-	-2.1
GROSS CONSUM	65.3	74.3	42.4	13.5	-	0.4	-	195.9
BUNKERS	-	2.3	-	-	-	-	-	2.3
INLAND CONSM	65.3	71.9	42.4	13.5	-	0.4	-	193.6
ELEC POWER S	46.2	5.9	0.6	13.5	0.0	-23.2	-	43.0
OTHER TRANSF	3.1	0.2	-2.6	-	-	-	-	0.7
ENERG INDUST	0.3	6.3	3.2	-	-	4.2	-	14.0
FINAL CONSUM	15.7	58.8	41.2	-	0.0	19.4	-	135.3
- NON ENERGY	0.5	7.9	-	-	-	-	-	8.4
- INDUSTRY	6.4	9.4	14.4*	-	-	6.7	-	36.9
- TRANSPORT	-	32.3	-	-	-	0.4	-	32.7
- HOUSEHOLD	8.5	9.0	25.2	-	0.0	12.4	-	55.2
STAT DIFFER	0.3	0.2	1.6	-	-	-	-	2.1

* Including non-energy consumption (+ 1 million toe).

ENERGY SUPPLY / DEMAND BALANCE - 1990   **UNITED KINGDOM**

million toe

	SOLID FUEL	PETROL PRODCT	GAS	NUCLR ENERGY	HEAT	HYDRO+ OTHERS	RENEW ENER	TOTAL
PRIM PRODUCT	70.0	90.0	35.0	20.5	-	0.4	-	215.9
TOT IMPORTS	4.5	54.6	12.1	-	-	-	-	71.2
TOT EXPORTS	2.5	68.0	-	-	-	-	-	70.5
STOCK CHANGE	-	-	-	-	-	-	-	-
GROSS CONSUM	72.0	76.6	47.1	20.5	-	0.4	-	216.6
BUNKERS	-	2.4	-	-	-	-	-	2.4
INLAND CONSM	72.0	74.2	47.1	20.5	-	0.4	-	214.2
ELEC POWER S	55.2	3.6	1.1	20.5	-1.2	-26.8	-	52.4
OTHER TRANSF	2.6	0.6	-1.6	-	-	-	-	1.6
ENERG INDUST	0.4	6.1	2.5	-	-	4.9	-	13.9
FINAL CONSUM	13.8	63.9	45.1	-	1.2	22.3	-	146.3
- NON ENERGY	0.3	8.5	1.0	-	-	-	-	9.8
- INDUSTRY	7.4	12.0	15.1	-	0.4	8.4	-	43.3
- TRANSPORT	-	35.5	-	-	-	0.6	-	36.1
- HOUSEHOLD	6.1	7.9	29.0	-	0.8	13.3	-	57.1

ENERGY SUPPLY / DEMAND BALANCE - 1995

million toe

	SOLID FUEL	PETROL PRODCT	GAS	NUCLR ENERGY	HEAT	HYDRO+ OTHERS	RENEW ENER	TOTAL
PRIM PRODUCT	70.0	90.0	37.0	22.5	-	0.5	-	220.0
TOT IMPORTS	5.9	55.8	13.7	-	-	-	-	75.4
TOT EXPORTS	2.5	66.5	-	-	-	-	-	69.0
STOCK CHANGE	-	-	-	-	-	-	-	-
GROSS CONSUM	73.4	79.3	50.7	22.5	-	0.5	-	226.4
BUNKERS	-	2.4	-	-	-	-	-	2.4
INLAND CONSM	73.4	76.9	50.7	22.5	-	0.5	-	224.0
ELEC POWER S	57.2	4.6	2.5	22.5	-1.6	-29.0	-	56.2
OTHER TRANSF	2.7	0.5	-1.5	-	-	-	-	1.7
ENERG INDUST	0.4	6.1	2.6	-	-	5.2	-	14.3
FINAL CONSUM	13.1	65.7	47.1	-	1.6	24.3	-	151.8
- NON ENERGY	0.3	10.2	1.0	-	-	-	-	11.5
- INDUSTRY	7.9	11.3	16.1	-	0.6	9.2	-	45.1
- TRANSPORT	-	37.0	-	-	-	0.6	-	37.6
- HOUSEHOLD	4.9	7.2	30.0	-	1.0	14.5	-	57.6

ENERGY SUPPLY / DEMAND BALANCE - 2000

million toe

	SOLID FUEL	PETROL PRODCT	GAS	NUCLR ENERGY	HEAT	HYDRO+ OTHERS	RENEW ENER	TOTAL
PRIM PRODUCT	70.0	88.0	38.0	34.8	-	0.5	0.2	231.5
TOT IMPORTS	2.9	59.7	12.7	-	-	-	-	75.3
TOT EXPORTS	2.5	66.5	-	-	-	-	-	69.0
STOCK CHANGE	-	-	-	-	-	-	-	-
GROSS CONSUM	70.4	81.2	50.7	34.8	-	0.5	0.2	237.8
BUNKERS	-	2.4	-	-	-	-	-	2.4
INLAND CONSM	70.4	78.8	50.7	34.8	-	0.5	0.2	235.4
ELEC POWER S	53.7	4.8	0.5	34.8	-1.9	-31.6	-	60.3
OTHER TRANSF	2.8	0.6	-1.3	-	-	-	-	2.1
ENERG INDUST	0.4	6.2	2.7	-	-	5.7	-	15.0
FINAL CONSUM	13.5	67.2	48.8	-	1.9	26.4	0.2	158.0
- NON ENERGY	0.4	11.4	1.0	-	-	-	-	12.8
- INDUSTRY	9.1	9.9	17.2	-	0.8	10.0	-	47.0
- TRANSPORT	-	38.9	-	-	-	0.6	-	39.5
- HOUSEHOLD	4.0	7.0	30.6	-	1.1	15.8	0.2	58.7

Summarized Energy Balance – UNITED KINGDOM

Table 3 p.203

in million toe	1973[a]	1980[a]	1983[a]	1990[b]	1995[b]	2000[b]
I. Gross Energy Consumption	227.04	202.28	195.97	216.6	226.4	237.8
– Bunkers	5.31	2.39	1.98	2.4	2.4	2.4
– Inland consumption	221.73	199.89	193.99	214.2	224.0	235.4
II. Inland Energy Consumption	221.73	199.89	193.99	214.2	224.0	235.4
– Solid fuels	79.17	69.88	65.39	72.0	73.4	70.4
– Oil	108.24	79.37	72.29	74.2	76.9	78.8
– Gas	25.11	39.89	42.38	47.1	50.7	50.7
– Primary electricity, etc.	9.21	10.75	13.93	20.9	23.0	35.5
III. Indigenous Production[1]	113.00	196.07	232.94	215.9	220.0	231.5
– Hard coal	78.67	74.73	69.24	70.0	70.0	70.0
– Lignite & peat	–	–	–	–	–	–
– Oil	0.68	79.70	117.02	90.0	90.0	88.0
– Natural gas	24.44	30.89	32.75	35.0	37.0	38.0
– Nuclear energy	8.88	10.41	13.54	20.5	22.5	34.8
– Hydro & geothermal[2]	0.33	0.34	0.39	0.4	0.5	0.5
– Others & renewables	–	–	–	–	–	0.2
IV. Net Imports[3]	112.88	12.72	-34.85	0.7	6.4	6.3
– Solid fuels	-0.85	1.77	-0.76	2.0	3.4	0.4
– Oil	113.06	1.95	-43.72	-13.4	-10.7	-6.8
– Natural gas	0.67	9.00	9.63	12.1	13.7	12.7
– Electricity[2]	0.00	0.00	–	–	–	–
V. Stock Changes[4]	-1.16	6.51	+2.13	–	–	–
– Solid fuels	-1.35	6.62	+3.10	–	–	–
– Oil	+0.19	-0.11	-0.97	–	–	–
– Gas	–	–	–	–	–	–
VI. Electricity Generation Input	72.66	69.79	66.64	81.0	87.3	94.3
– Solid fuels[5]	44.17	50.14	46.42	56.3	58.1	54.1
– Oil	18.34	8.14	5.93	3.6	4.6	4.8
– Natural gas	0.94	0.76	0.36	0.2	1.6	0.1
– Nuclear energy	8.88	10.41	13.54	20.5	22.5	34.8
– Hydro & geothermal[2]	0.33	0.34	0.39	0.4	0.5	0.5
– Others & renewables	–	–	–	–	–	–

Main indicators (related to long term objectives)

	1973–1963	1980–1975	1983–1980	1990–1983	2000–1990
Inland energy annual growth rate	+1.8%	-0.1%	-1.0%	+1.4%	+0.9%
GDP annual growth rates	+3.2%	+1.5%	+1.3%	+2.2%	+2.5%
Energy-GDP ratio	0.56	-0.07	-0.77	0.65	0.38

	1973	1980	1983	1990	1995	2000
Share of oil in gross energy consumption	50.0%	40.4%	37.9%	35.4%	35.0%	34.1%
Share of coal and nuclear in in electricity production	73.0%	86.8%	90.0%	94.8%	92.3%	94.3%
Supply dependance on imports	50.0%	6.3%	-17.8%	+0.0%	2.8%	2.6%

Sources: a. Statistical Office of the European Communities
        b. "Energy 2000"

Notes: 1. Production of primary sources, including recovered products.
        2. The conversion of electricity, including hydro and geothermal, is based on its actual energy content: 3600 kjoules/kWh or 860 kcal/kWh
        3. The (-) sign means net exports
        4. The (-) sign means a stock decrease
        5. Including coke oven gas and blast furnace gas (derived from coal)

CHAPTER V

THE COMMUNITY'S PLACE IN THE WORLD ENERGY MARKETS

A. INTRODUCTION

1. The Community is the world's leading importer of energy, and will probably still be in that position in the year 2000. Any forward study of the Community energy market must therefore inevitably be set in a world context, with a review of the energy supply and demand outlook in other regions of the world.

2. In 1983 the Community's gross primary energy consumption amounted to 907 million toe or 12.5% of world consumption while the Community's population was just under 6% that of the world total. So the Community was consuming 3.3 toe per capita per annum. By comparison, per capita energy consumption in 1983 amounted to 7.5 toe in the USA, 4.5 toe in the USSR, 2.7 toe in Japan, 1.9 toe in Mexico, 0.6 toe in China and 0.3 toe in India.

   In the same year primary energy production in the Community reached 516 million toe or 7% of world energy production. The Community thus depended on other regions of the world for 42% of its energy supplies. During the same period, Japan imported 85% of its energy requirements and the USA 11%, while the USSR exported 17% of its net energy production, Mexico 38% and Saudi Arabia 91%.

3. Up to the end of the century, the relative energy positions of the major world economic groups are likely to be maintained. In 2000 the Community's consumption will represent no more than 10% of world demand for energy but it will depend on external supplies for nearly 45% of its energy requirements.

   The global figures set out here are obtained from an analysis of possible trends in the conditions of energy supply and demand in the principal countries and regions of the world up to the year 2000. A central projection, based on a set of assumptions considered to be sufficiently probable, forms the reference projection. The principal uncertainties attached to the assumptions adopted are also identified together with possible deviation spreads for world energy supply and demand. Equilibrium on the world energy market in the year 2000 should lie somewhere within the outer limits of these deviations.

B.   WORLD ENERGY CONSUMPTION: REFERENCE PROJECTION

4.   An initial set of assumptions consistent with those described for
the Community itself was used to produce a reference projection for
world energy demand.[1]   The global results are presented in the
table below.   More detailed data showing the position of the
Community and the other major regions of the world are annexed.

Average annual rate of change in gross energy consumption[2]

	1980/1970	1983/1980	1990/1983	2000/1990
Industrialized countries	+2.2%	−2.2%	+2.5%	+1.4%
Developing countries	+3.1%	+2.7%	+4.2%	+2.9%
Centrally planned economies	+5.4%	+1.1%	+2.6%	+2.5%
World	+3.3%	−0.3%	+2.8%	+2.0%

This reference projection reflects two essential trends:

- a gradual and generalized slowdown in the growth of energy
  requirements as new technologies penetrate the consumer markets
  and lead to a more rational use of energy, **including in
  developing countries after 1990;**

- the **increasing share of developing countries and planned
  economies in the structure of world energy consumption.**

5.   The reference projection assumes a world average economic growth
rate of around 3.3% over the period 1983-2000: nearly 3% in
industrialized countries and over 4% in developing countries. These
rates are within the spread of all the long-term projections
published in 1983 and 1984.

Among the industrialized countries, the economic growth rate of the
Community is projected at 2.6%, the United States 2.8%, the other
industrialized countries nearly 3% and Japan just over 3%. The
corresponding growth in primary energy demand will average 1.8% to
2% per annum with a more marked decoupling of energy demand from
economic growth in the Community and the United States than in the
other industrialized countries.

---

[1] This conspectus of world energy prospects up to the end of the century was
primarily based on 12 important studies carried out from 1981 to 1983:
Chevron (1982), Conoco (1983), Exxon (1981), Shell (1982), Petroleum
Economics Limited (1983), Chase Manhattan Bank (1983), California Energy
Commission (1983), IEA (1982), IIASA (1983), the 11th World Oil Congress
(RTD 9, 1983), the 12th Congress of the World Energy Conference (report by
J.R. Frisch, 1983) and IFP (1983).

[2] Inland consumption and bunkers (which account for 3.5% of the oil
consumption of the industrialized countries).

6. In comparison with previous forecasts, the projected average annual rate of growth of world demand for energy may appear low at 2.4%, but the projection seems justified for several complementary reasons.

    In the industrialized countries, the industrial restructuring which started in the mid 1970s, the gradual penetration of new energy-saving technologies and the widespread improvement in efficiency of use among consumers all will tend to slow down the growth in energy requirements.

    In countries with centrally planned economies, a more rational use of energy has also become at least more of a priority, especially since prices for transferring energy between these countries have been increasingly aligned on those prevailing on the international market.

    In non-oil developing countries, growth in the use of imported commercial energy may be held down by foreign exchange constraints, the debt burden, on the one hand, and efforts to improve energy planning and the promotion of less energy-intensive industrial growth, on the other.

7. The assumed structure of world demand by form of energy in the year 2000 (reference projection) is summarized in the table below. The same information covering several years (1970, 1980, 1983, 1990, 2000) and a greater number of geographical regions (Community, USA, Japan, other industrialized countries, OAPEC, OLADE, other developing countries, USSR, China, other centrally planned economies) is given in the tables annexed.

World gross primary energy consumption

	Solid fuels	Oil	Natural gas	Nuclear power	Hydro	Other[3]	Total
						million toe	
1983 situation							
Industrialized countries	949	1642	695	200	106	2	3594
Developing countries	109	535	155	7	43	404	1253
Centrally planned economies	1171	629	486	33	31	75	2425
World	2229	2806	1336	240	180	481	7272
Projection for 2000							
Industrialized countries	1530	1677	847	589	148	83	4875
Developing countries	238	919	329	38	88	621	2233
Centrally planned economies	1685	840	887	147	46	110	3715
World	3453	3436	2063	774	282	814	10822

Average annual % change over the period 1983-2000

	Solid fuels	Oil	Natural gas	Nuclear power	Hydro	Other	Total
Industrialized countries	+2.8%	+0.1%	+1.2%	+6.6%	+1.8%	+19.3%	+1.8%
Developing countries	+4.7%	+3.2%	+4.4%	+8.5%	+4.9%	+2.5%	+3.4%
Centrally planned economies	+2.2%	+1.7%	+3.6%	+9.9%	+3.1%	+1.9%	+2.5%
World	+2.6%	+1.2%	+2.6%	+7.2%	+2.8%	+3.0%	+2.4%

Consumption of all forms of primary energy should increase between 1983 and 2000 in all regions, except for oil consumption in the industrialized countries, which will remain more or less stable throughout the period. The relative share of oil in meeting energy requirements will decrease in all regions, however, as a result of substitution efforts. Natural gas will cover a growing share of requirements, except in the industrialized countries. The shares of solid fuels and nuclear energy will increase especially in the industrialized countries and centrally planned economies.

Non-commercial energy sources should also lose a little of their consumption share in the developing countries, as some of them become increasingly concerned about advancing desertification caused by massive use of firewood without any adequate reafforestation policy. But there is major uncertainty about the potential contribution of these forms of energy.

---

[3] The energy sources included in this column are not generally the subject of international trade, and production and consumption balance out in the large regions. In the industrialized countries they are new and renewable energy sources; the figures for the developing countries include non-commercial energy sources.

WORLD GROSS ENERGY CONSUMPTION ✳
(by regions)

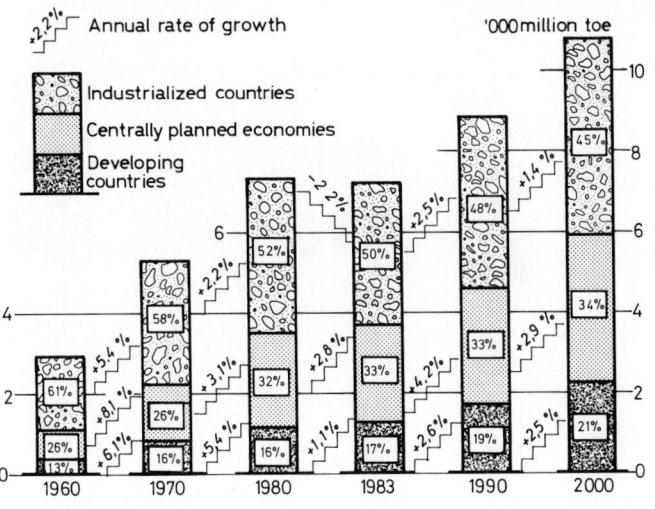

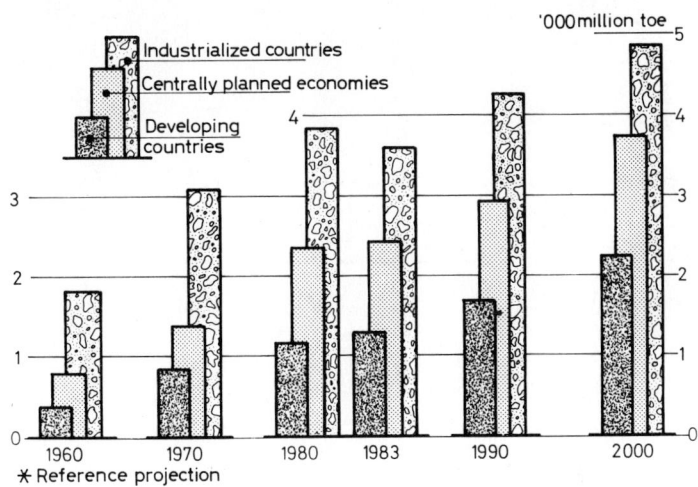

✳ Reference projection

8. The additional demand for energy generated if the economy grew
    faster than assumed in the reference projection would have to be
    met largely by increased consumption of hydrocarbons, mainly oil.

WORLD GROSS ENERGY CONSUMPTION ✳
(by sources)

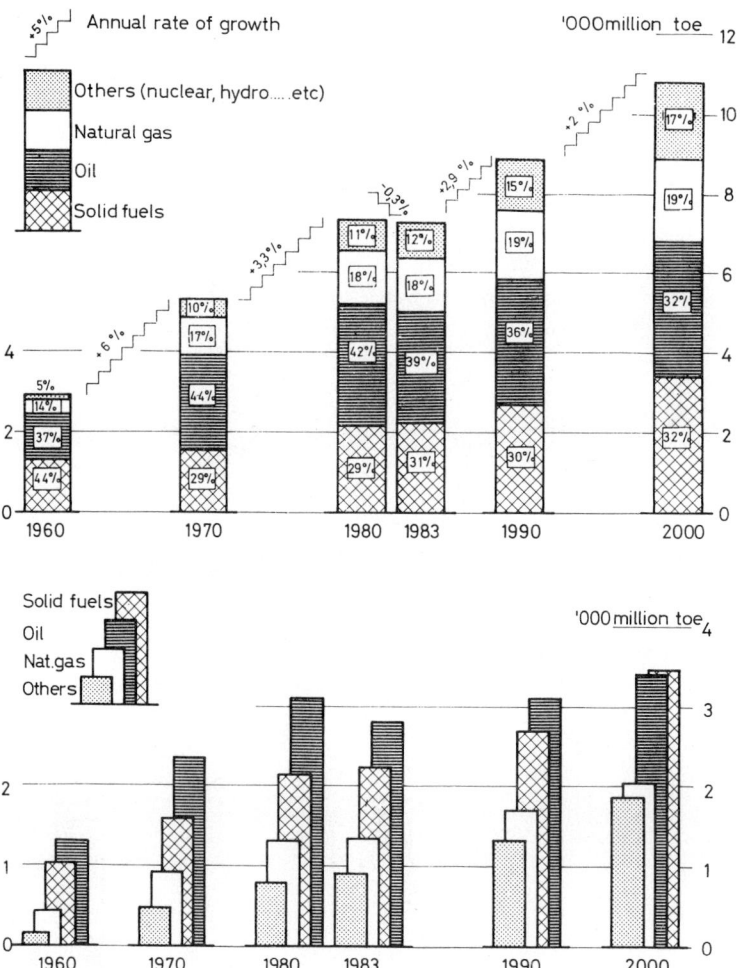

✳ Reference projection

It should be noted that although the oil consumption figure adopted in the reference projection (3 400 million toe) is at the bottom end of the spread given by all the world projections listed in footnote 1 on page 217 (Chapter V), the level at the top of the spread is nearly 4 200 million toe.

Oil's share of world energy consumption can therefore be expected to lie between 32% and 35% in 2000, compared with 39% today.

In 2000 the industrialized countries will probably account for about 48% of world oil consumption, compared with 58% today and 68% in 1970. This downward trend could be stronger still if the

industrialized countries continue to switch to natural gas.   That
will depend, however, on how much extra natural gas can be put on
the market at competitive prices.

C.   WORLD ENERGY PRODUCTION: REFERENCE PROJECTION

9.   The reference projection underlines the need for a general increase
     in the production of all primary energy sources up to the end of
     the century.   The broad trends of world energy production will not
     radically depart from the historical trend because:

     - in 2000 the industrialized countries will produce enough to cover
       only 75% of their energy requirements, the balance coming
       essentially from developing countries producing hydrocarbons;

     - the developing countries will continue to supply nearly 60% of
       world oil production, exporting about half their output;

     - the centrally planned economies will remain self-sufficient and
       could even be net exporters of solid fuels and natural gas.

     The reference projection for world production of primary energy is
     summarized in the table below. More detailed tables are annexed.

World primary energy production

	Solid fuels	Oil	Natural gas	Nuclear	Hydro	Others[4]	Total
Situation 1983:							
Industrialized countries	956	765	637	200	106	2	2666
Developing countries	91	1252	199	7	43	404	1996
Centrally planned economies	1178	750	516	33	31	75	2583
World	2225	2767	1352	240	180	481	7245
Projections for 2000:							
Industrialized countries	1578	671	651	589	148	83	3721
Developing countries	158	1925	431	38	88	621	3261
Centrally planned economies	1720	825	1005	147	46	110	3853
World	3456	3421	2087	774	282	814	10835

million toe

(a) Solid fuels

10.   A major increase in the production of solid fuels is forecast in the industrialized countries (+65%) and the centrally planned economies (+45%) between 1983 and 2000. In the industrialized countries the biggest increases will be in the USA (+400 million toe), Australia (+100 million toe), Canada (+60 million toe) and South Africa (+30 million toe).

This trend is conditional upon a true international market in coal being established in which prices will reflect costs rather than be aligned on the guide price of a competing source of energy. The production level assumed also implies a return to coal for certain industrial uses in the industrialized countries, on the supposition that the promotional policies followed in this sector will be successful.

In the centrally planned economies, the increase in production will come mainly from China (+350 million toe), which will step up output to meet its own needs, the USSR (+70 million toe), and the countries of Eastern Europe (+70 million toe). This additional output will allow a slight increase in exports.

Substantial increases in coal output, primarily for inland consumption, can also be expected in some developing countries such as India and South Korea.

(b) Oil

11.   In the reference projection, oil production is assumed to stagnate or even decline slightly between 1983 and 2000 in all industrialized countries except Canada.

---

[4] The energy sources included in this column are not generally the subject of international trade, and production and consumption balance out in the large regions. In the industrialized countries they are new and renewable sources; the figures for the developing countries include non-commercial energy sources.

This is because proven deposits already in production or soon to come on stream will reach their maximum level of output before the end of the century or will have already started to decline. It is most unlikely that the conceivable discovery of new giant fields will substantially increase the output of the industrialized countries before 2000, bearing in mind the lead times for production and transport infrastructure in the areas now being prospected, which are particularly difficult.

On the other hand, it cannot be ruled out that a sizeable increase in the industrialized countries' oil production may come from the rapid generalization after 1990 of tertiary recovery methods which would increase the rate of production in existing fields.

Nor do the countries with centrally planned economies, apart from China, seem to be in a position to step up their oil production substantially. In China annual production could increase by around 50 million toe. This extra oil will not necessarily be consumed by the domestic market, for major energy-saving and substitution efforts may be deployed in order to release for export as much oil as possible so as to earn sufficient foreign currency to finance the purchase of industrial and high-technology products from the industrialized countries.

12. In order to obtain a clearer picture of oil production trends in the developing countries, we have to identify those among them which are primarily consuming countries, those which are broadly self-sufficient and those which are traditionally net exporters (among which are the countries of the Gulf).

In the reference projection, oil production in the OAPEC countries[5] would have to reach 1 000 million toe in 2000 in order to ensure that world supply and demand are in equilibrium. This level is 70% up on 1983, but only represents a return to OAPEC's 1980 production levels. Against this background, the OAPEC countries would produce approximately 80% of their estimated operational capacity[6] of 24.5 million barrels per day.

Saudi Arabia could supply half of this output and Iraq approximately one fifth. According to the reference projection, this level of output will be translated into net exports by the OAPEC countries in 2000 of approximately 900 million toe.

13. The countries of Latin America should return to their 1983 level of output in the year 2000, i.e. 330 million tonnes, after peaking 10% higher in 1990. Unlike the current situation, however, this output will not leave much surplus for export, owing to the increase in inland oil demand in the OLADE countries.[7]

----

[5] OAPEC: Organization of Arab Petroleum Exporting Countries (Algeria, Bahrein, Egypt, Iraq, Kuwait, Libya, Oman, Qatar, Saudi Arabia, Syria, Tunisia, United Arab Emirates).
[6] Defined by the IEA as "maximum sustainable capacity".
[7] OLADE: Latin-American Energy Organization, whose members are Mexico and the countries of Central America, the Caribbean and South America (such as Venezuela, Argentina and Brazil).

These countries include two major oil producers:

- Venezuela, whose output should be about 100 million toe per annum throughout the period;

- Mexico, whose fast-rising output could approach 200 million toe in 1990; beyond that date, the current controversy about the real extent of Mexico's reserves has prompted the cautious estimate of a level of output of less than 150 million toe in 2000. If the most optimistic estimates of Mexican oil resources are confirmed, the country's oil production could rise to 250 million toe at the end of the century.

Three other OLADE countries (Colombia, Ecuador and Peru) are also oil producers. Their total production level should increase by a third between 1983 and 2000 to 40 million toe, supporting net exports estimated at 5 million to 10 million toe per annum.

14.   Certain other developing countries should be treated separately in view of their importance on the oil market.

In Africa, Nigeria is and will probably remain the principal oil producer. The country's output could level out in 1990 at 75 million toe, i.e. 90% of the operational production capacity planned for the end of the next decade. Five other African producers will assume increasing importance between now and the end of the century: Angola, Gabon, Cameroon, Congo and Zaire. Together they will produce 50 million toe of oil in 2000. After satisfying their own domestic needs, they should have a surplus of around 40 million toe of oil for export.

WORLD SUPPLY AND DEMAND OF HYDROCARBONS
(UNITS: MILLION TOE)

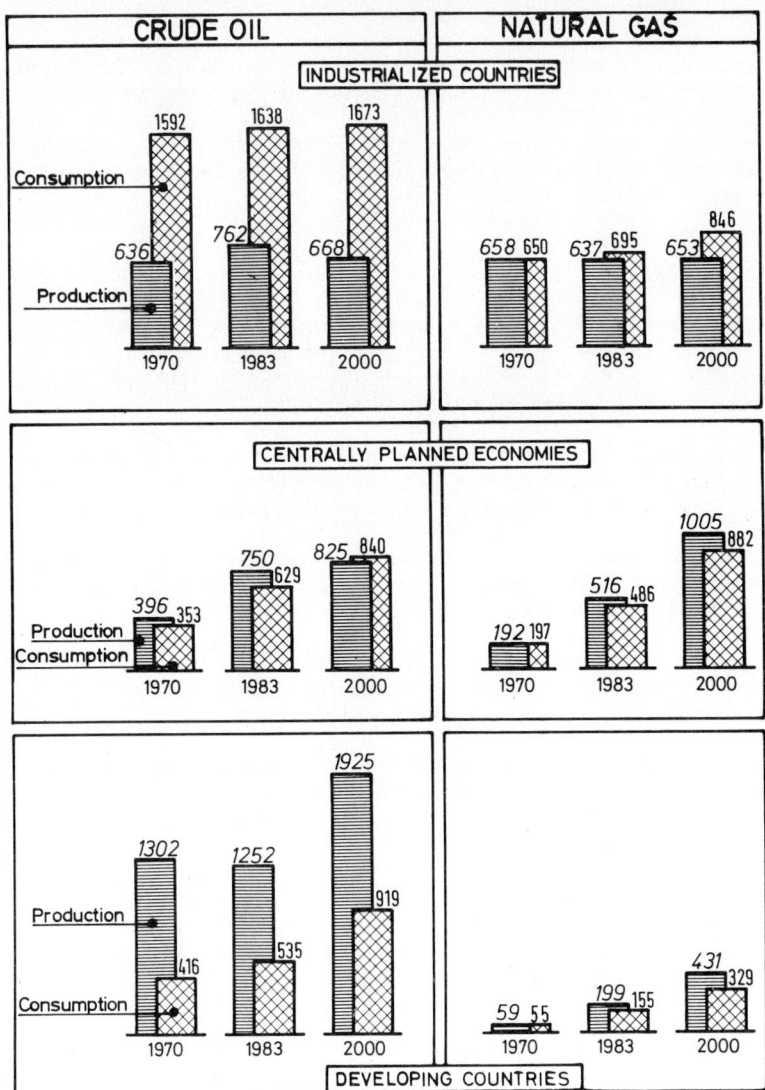

CRUDE OIL	NATURAL GAS

INDUSTRIALIZED COUNTRIES

Consumption: 1592 (1970), 1638 (1983), 1673 (2000)
Production: 636 (1970), 762 (1983), 668 (2000)

Natural gas: 658 650 (1970), 637 695 (1983), 653 846 (2000)

CENTRALLY PLANNED ECONOMIES

Production / Consumption: 396 353 (1970), 750 629 (1983), 825 840 (2000)

Natural gas: 192 197 (1970), 516 486 (1983), 1005 882 (2000)

Production: 1302 (1970), 1252 (1983), 1925 (2000)
Consumption: 416 (1970), 535 (1983), 919 (2000)

Natural gas: 59 55 (1970), 199 155 (1983), 431 329 (2000)

DEVELOPING COUNTRIES

ıwo Asian producers, Iran and Indonesia, have substantial export capacities:

- Iran's output could double between 1983 and 2000 to reach a peak of 250 million toe using 80% of current installed capacity. Such a level of output presupposes a return to political stability and the recovery of oil production. This assumption is not unreasonable in the medium term in view of Iran's increasing financial requirements. If these assumptions are disproved in the event, Iran's maximum output would be at least 100 million toe lower than the peak figure given above.

- Indonesia's output, is assumed to return to its 1980 level, i.e. 80 million toe, of which half would be for export.

Two other Asian countries, India and Malaysia, are major oil producers, but their output is expected to be absorbed entirely by increasing domestic needs.

(c) <u>Natural Gas</u>

15. In contrast to oil, where uncertainties about potential levels of output often come down to political considerations, there is a more general uncertainty about potential world output of natural gas. This uncertainty is fed by three major factors:

- what proportion of the known new deposits of natural gas in the industrialized countries will be brought on stream before the end of the century, and under what conditions?

- how probable is the creation, by the year 2000, of a world LNG market which will make it possible to exploit the gas associated with oil production, in particular in the Gulf?

- how far advanced will be the projects to market the gas resources of the USSR on both the home and the export markets?

16. Although it is difficult to answer these questions, assumptions have been made for the reference projection. They are guided by two criteria: whether or not the projects have been started, and their probable cost.

The assumptions are that the production of natural gas will tend to fall after 1990 in the industrialized countries, rise by around 50% between 1983 and 2000 in the developing countries and double in the centrally planned economies.

In the industrialized countries, no significant development of huge new giant fields is posited by the year 2000, when the impact of the Norwegian Troll field is assumed to be only marginal. Domestic production of natural gas would thus decrease appreciably after 1990 in the Community, the USA and the other OECD countries, except Canada, where a natural gas export capacity of around 30 million toe could be available at the end of the century.

Any increase in the production potential of countries with centrally planned economies hinges almost exclusively on developments in the USSR. Despite a substantial increase in its inland consumption, the Soviet Union may be exporting more than 200 million toe of natural gas in 2000, of which only half would go to East European countries. This very probable trend would mean an appreciable increase in Soviet exports of natural gas to other European countries.

The developing countries' natural gas output could more than double between 1983 and 2000, allowing them to export approximately 100 million toe to the industrialized countries in 2000.

- In the OLADE countries, the increase in production will be absorbed entirely by the growth in domestic demand.

- In Africa and Asia, natural gas production will generally match the increased production of oil and expand at the same rate. Two countries, Nigeria and Iran, could record a higher increase; in view of the reserves available, their natural gas output in 2000 could be at least quintupled over the period, to 25 million and 50 million toe respectively. This result is hedged about, however, by political and economic uncertainties similar to those relating, for example, to Iranian oil.

- The OAPEC countries will also substantially increase their production of natural gas, up to 2.5 times its current volume by 2000. The bulk of the increase will be used primarily to meet home market demand, limiting these countries' export capacity to approximately 30 million toe.

The production levels assumed both for the OAPEC countries and for Nigeria and Iran, though showing marked growth, are far short of production potential, particularly for gas associated with oil.

(d) Nuclear energy

17. Nuclear-generated electricity should grow rapidly up to the end of the century in all world regions.

From the point of view of energy substitution, however, nuclear power will play an important role only in the industrialized countries and in Eastern Europe, where installed capacities could reach 375-400 GWe and 100-125 GWe respectively.

In the developing countries, only such countries as India, South Korea, Taiwan, Pakistan, Mexico, Argentina and Brazil have nuclear installations already operational or under construction or a large enough production capacity to warrant building nuclear stations. Installed nuclear capacity in the developing countries should be between 20 and 30 GWe in 2000.

Two major uncertainties hang over the future expansion of nuclear energy in the world: the rate of penetration of nuclear technologies for peaceful purposes in China, and the importance that this type of electricity generation could have for certain countries in the Near and Middle East.

(e) Hydro power and other sources of energy

18. Hydroelectricity generation should rise steadily in all parts of the world between now and the year 2000.

Growth could be more marked in certain developing countries in Africa and Asia where the size of a country, its industrial electricity requirements (particularly industries carrying out-c primary processing of ores) and suitable geographical sites will make the development of this form of electricity generation particularly attractive.

19. The other sources of energy classified under "other" in the foregoing tables are quite different in the industrialized countries as against the developing countries.

In the industrial countries, the emphasis is on new technologies to exploit renewable sources (wind, solar, biomass etc.), which should penetrate markets quite rapidly but meet only an insignificant share of total requirements.

In the developing countries, these "other" energy sources are mainly non-commercial and their widespread use is often contrary to the long-term interest of countries anxious to ensure sound management of their natural resources. These non-commercial energy sources covered 20% of the developing world's energy requirements in 1983.

Since most developing countries are now alive to the problem of desertification caused by unregulated deforestation, it is assumed that the use of non-commercial forms of energy will be discouraged as much as possible and that their share in meeting the energy requirements of these countries will gradually decline.

D. THE UNCERTAINTIES ON THE WORLD ENERGY MARKET

20. The reference projection shows one possible outlook for the world energy supply and demand balance. The reality could differ in many ways, depending on the many uncertainties in the assumptions on which the projection is based.

By considering possible variations in the assumptions we have attempted to define the margins of deviation within which various possible alternative projections should reasonably lie.

(a) Energy demand

21. The first source of uncertainty is about rates of economic growth. A rate of 3.3% may prove too pessimistic. An additional increase in economic growth of say 0.5% would trigger a corresponding increase in world demand for energy of 500 million toe up to the year 2000. This additional requirement would have to be largely covered by hydrocarbons.

The price of energy constitutes a second uncertainty.

A rising trend in energy prices has a depressive effect on the economy but it encourages more efficient energy use, and both these factors tend to hold down demand for energy.

If over the coming years energy prices are much lower than assumed in the reference projection, on the other hand, the likelihood is that in the medium term there will be an upturn in energy demand and a contraction of potential future supply. These two divergent trends contain the seeds of a sudden jump in prices at some future date.

The effect of high or low energy prices would be a deviation in the world energy demand in 2000 from its reference level of -700 million to +600 million toe.

A third uncertainty stems from the rate of penetration of new energy-efficient technologies on consumer markets and the rate of transfer to developing countries of the expertise and know-how involved. A less favourable trend than that assumed for the reference projection would lead to an estimated additional consumption of energy in 2000 of 400 million toe (assuming the same rate of economic growth as in the reference scenario).

A fourth source of uncertainty is the scope for further substitution between forms of energy, and more especially between fossil fuels, assuming that price relativities do not change appreciably.

There seems to be little scope by 2000 for any further significant shift to coal, which is used chiefly in specific markets and which is not the most convenient fuel for industrial purposes.

For natural gas, the situation could change significantly, depending on supply conditions and especially on decisions whether or not to bring new gas fields on stream.

(b) Energy supply

22. Uncertainties relate in the main to output of hydrocarbons.

If one compares demand and supply of oil up to the year 2000, it is clear that market equilibrium will continue to depend primarily on the level of OPEC exports. This is mainly a political uncertainty, as the reserves necessary to maintain the desired production levels are known and adequate. In the reference projection, OPEC's oil production is put at about 29.5 million barrels a day, i.e. nearly 85% of their maximum operational production capacity in 2000. If necessary, and if they consider it politically desirable, these countries could market additional quantities of oil amounting to as much as 5 million barrels a day.

Mexico could produce a further 2 million barrels a day if the optimistic assumption for this country's reserves proves to be correct. An additional output of 1 million barrels a day could also come from the industrialized countries.

Available natural gas supplies constitute another major source of uncertainty in the period up to 2000. The reference projection allows for a balance between world demand and supply of gas at a level somewhat above 2 000 million toe. If new gas fields come on stream in the industrialized countries before the end of the century, say in North America and Europe, an extra 300 million toe of natural gas would be available (a third of which in Europe). An extraction potential higher than the level of output assumed in the reference projection also exists in certain developing countries. The resulting additional export potential   would be around 200 million toe, a third of which would come from Africa, a third from the Gulf, 20% from Latin America and the rest from the Far East.

There are also uncertainties, though of less crucial impact, about the contribution of other forms of energy in meeting world energy requirements:

- **solid fuels**: the exploitation of new mining regions long term will depend both on an expanding international market and on the evolution of prices;

- **nuclear energy**: its contribution in 2000 could be affected, as it was in the past, by an accident like that at Three Mile Island;

- **new and renewable energy sources**: their rate of penetration into the energy system is subject to many uncertainties of a technological and economic nature.

E. BALANCING WORLD ENERGY DEMAND AND SUPPLY IN THE YEAR 2000

23. The table below summarizes and quantifies, for the world as a whole, the principal uncertainties affecting the world energy market in 2000 as compared with the reference projection.

Energy demand and supply in the world in 2000
('000 million toe)

DEMAND		SUPPLY	
Reference projection	10.8	Reference projection	10.8
Uncertainties:		Uncertainties:	
- economic growth	-0.5/+0.5	- oil production	-0.6/+0.5
- energy prices	-0.7/+0.6	- natural gas production	-0.3/+0.5
- efficiency of use	0/+0.4	- other forms of energy	-0.3/+0.3
Total	9.6/12.2	Total	9.6/12.1
of which: oil	3.0/4.2	of which: oil	2.8/3.9
natural gas	1.8/2.5	natural gas	1.8/2.6
other sources	4.8/5.5	other sources	5.0/5.6

This table shows that the main potential field of imbalance is the world oil market.

There should be no problem regarding other sources, since the potential for coal production should make it possible to adapt supply to demand requirements, either by directly substituting coal for non-commercial energy sources, or by making greater use of it in electricity generation or by allowing it greater penetration into industrial markets.

The potential imbalance in oil, however, is a more difficult matter. Even if as a matter of inevitability equilibrium is maintained in the long run through price adjustments in the medium term there could be serious potential disequilibria.

WORLD ENERGY BALANCE
Uncertainties in 2000

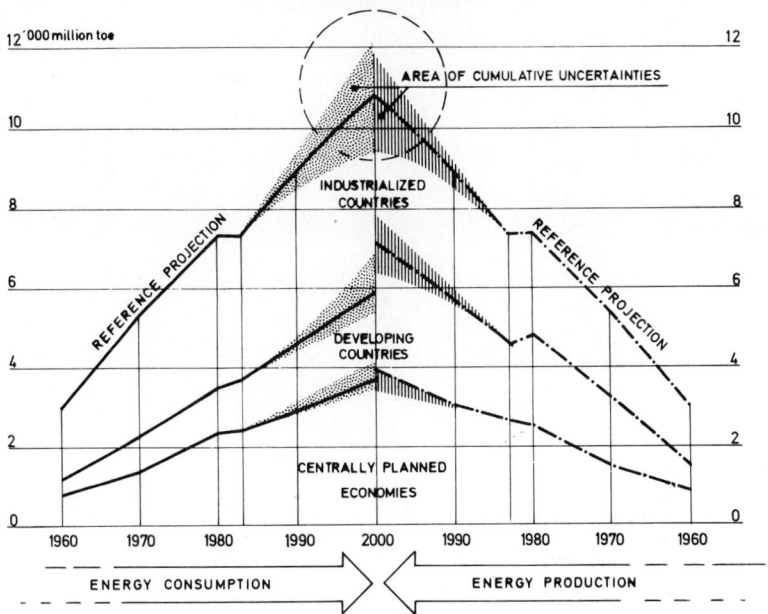

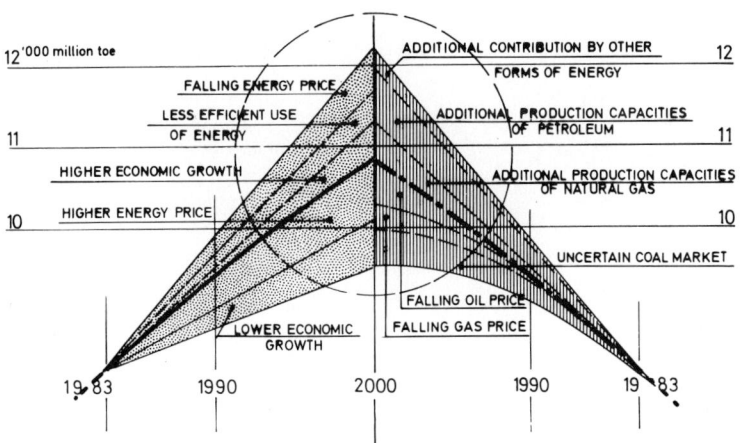

Sustained low oil prices, combined with higher rates of world economic growth than assumed in the reference case, would undoubtedly put pressure on world oil supplies over time as oil demand increased and the development of oil (and gas) from higher cost sources slowed down. This would in turn create the conditions for renewed upward price pressure.

*

*   *

ANNEX: World energy balances by major geographical region

TABLE 1: WORLD ENERGY BALANCE

1970

A. PRIMARY ENERGY PRODUCTION

Million toe

	Solid fuels	Oil	Natur. gas	Nuclear energy	Hydro electri	other (1)	Total
EEC(a)	224	14	60	13	13	1	325
USA	390	509	501	5	22	–	1427
Japan	30	1	2	1	7	–	41
Other industrialized countries	79	114	95	1	36	–	325
OAPEC (b)	–	711	6	–	1	–	717
OLADE (c)	6	277	38	–	5	49	375
Other developing countries	55	315	15	–	6	255	646
USSR	310	353	157	1	11	–	832
China	212	24	3	–	1	52	292
Other countries with centrally planned economies	248	19	32	0	3	14	316
World (*)	1554	2337	909	21	105	371	5297

B. GROSS PRIMARY ENERGY CONSUMPTION (2)

Million toe

	Solid fuels	Oil	Natur. gas	Nuclear energy	Hydro electri	other (1)	Total
EEC(a)	257	515	60	13	13	1	859
USA	336	663	509	5	22	–	1535
Japan	64	199	3	1	7	–	274
Other industrialized countries	91	215	78	1	36	–	421
OAPEC (b)	1	18	4	–	1	–	24
OLADE (c)	8	110	36	–	5	49	208
Other developing countries	55	288	15	–	6	255	619
USSR	298	263	159	1	11	–	732
China	212	28	3	–	1	52	296
Other countries with centrally planned economies	240	62	35	0	3	14	354
World (*)	1562	2361	902	21	105	371	5322

Notes: (1) Industrialized countries: new and renewable energy sources
Developing countries: non-commercial energy sources.
(2) Inland consumption + bunkers
(a) European Economic Community (EUR-10)
(b) Organization of Arab Petroleum Exporting Countries
(c) Latin-American Energy Organization.
Methodology: The methodology adopted for the establishment of energy balance is that used by the Statistical Office of the European Communities.
(*) The discrepancy between world production and world consumption is explained by possible stock movements and statistical deviation.

TABLE 2: WORLD ENERGY BALANCE

1980

A. PRIMARY ENERGY PRODUCTION

Million toe

	Solid fuels	Oil	Natur. gas	Nuclear energy	Hydro electri	Other (1)	Total
EEC(a)	185	91	129	43	12	2	462
USA	541	494	457	60	24	−	1576
Japan	12	1	2	20	8	−	43
Other industrialized countries	194	134	103	20	51	−	502
OAPEC (b)	−	1006	54	−	2	−	1062
OLADE (c)	9	303	67	1	18	60	458
Other developing countries	75	325	43	4	14	327	788
USSR	330	606	385	20	22	−	1363
China	435	107	20	−	5	50	617
Other countries with centrally planned economies	370	19	27	6	2	13	437
World (*)	2151	3086	1287	174	158	452	7308

B. GROSS PRIMARY ENERGY CONSUMPTION (2)

Million toe

	Solid fuels	Oil	Natur. gas	Nuclear energy	Hydro- electri	Other (1)	Total
EEC(a)	223	520	169	43	12	2	969
USA	465	839	482	60	24	−	1870
Japan	61	252	23	20	8	−	364
Other industrialized countries	169	326	69	20	51	−	635
OAPEC (b)	1	66	44	−	2	−	113
OLADE (c)	14	196	60	1	18	60	349
Other developing countries	81	241	26	4	14	327	693
USSR	330	450	350	20	22	−	1172
China	435	95	20	−	5	50	605
Other countries with centrally planned economies	365	112	70	6	2	13	568
World (*)	2144	3097	1313	174	158	452	7338

Notes: (1) Industrialized countries: new and renewable energy sources
              Developing countries: non-commercial energy sources
       (2) Inland consumption + bunkers
       (a) European Economic Community (EUR-10)
       (b) Organization of Arab Petroleum Exporting Countries
       (c) Latin-American Energy Organization
Methodology: The methodology adopted for the establishment of energy balances is that
              used by the Statistical Office of the European Communities.
       (*) The discrepancy between world production and world consumption is explained
           by possible stock movements and statistical deviation.

TABLE 3: WORLD ENERGY BALANCE

1983

A. PRIMARY ENERGY PRODUCTION

Million toe

	Solid fuels	Oil	Natur. gas	Nuclear energy	Hydro-electri	Other (1)	Total
EEC(a)	174	132	120	76	12	2	516
USA	543	495	410	71	28	-	1547
Japan	10	1	2	25	8	-	46
Other industrialized countries	229	137	105	28	58	-	557
OAPEC (b)	-	580	59	-	2	-	641
OLADE (c)	11	335	86	1	24	62	519
Other developing countries	80	337	54	6	17	342	836
USSR	348	624	451	26	23	-	1472
China	445	106	16	-	5	60	632
Other countries with centrally planned economies	385	20	49	7	3	15	479
World (*)	2225	2767	1352	240	180	481	7245

B. GROSS PRIMARY ENERGY CONSUMPTION (2)

Million toe

	Solid fuels	Oil	Natur. gas	Nuclear energy	Hydro-electri	Other (1)	Total
EEC(a)	212	438	165	76	14	2	907
USA	493	726	429	71	28	-	1747
Japan	57	209	24	25	8	-	323
Other industrialized countries	187	269	77	28	56	-	617
OAPEC (b)	1	68	40	-	2	-	111
OLADE (c)	16	204	80	1	24	62	387
Other developing countries	92	263	35	6	17	342	755
USSR	348	435	385	26	23	-	1217
China	445	93	16	-	5	60	619
Other countries with centrally planned economies	378	101	85	7	3	15	589
World (*)	2229	2806	1336	240	180	481	7272

Notes: (1) Industrialized countries: new and renewable energy sources
           Developing countries: non-commercial energy sources
       (2) Inland consumption + bunkers
       (a) European Economic Community (EUR-10)
       (b) Organization of Arab Petroleum Exporting Countries
       (c) Latin-American Energy Organization.
Methodology: The methodology adopted for the establishment of energy balances is that
              used by the Statistical Office of the European Communities.
       (*) The discrepancy between world production and world consumption is explained
           by possible stock movements and statistical deviation.

TABLE 4: WORLD ENERGY BALANCE, MEDIUM-TERM PROJECTION

1990

A. PRIMARY ENERGY PRODUCTION

Million toe

	Solid fuels	Oil	Natur. natur.	Nuclear energy	Hydro- electri	Other (1)	Total
EEC(a)	175	111	115	145	13	4	563
USA	718	454	432	148	29	2	1783
Japan	10	2	–	49	9	8	78
Other industrialized countries	321	153	154	57	69	2	756
OAPEC (b)	–	873	88	–	2	4	967
OLADE (c)	14	371	113	4	33	73	608
Other developing countries	105	472	87	16	27	420	1127
USSR	375	625	625	45	25	–	1695
China	560	100	15	1	7	70	753
Other countries with centrally planned economies	415	18	55	17	4	18	527
World (*)	2693	3179	1684	482	218	601	8857

B. GROSS PRIMARY ENERGY CONSUMPTION (2)

Million toe

	Solid fuels	Oil	Natur. gas	Nuclear energy	Hydro electri	Other (1)	Total
EEC(a)	242	441	190	145	12	4	1034
USA	633	753	484	148	29	2	2049
Japan	71	219	49	49	9	8	405
Other industrialized countries	236	295	109	57	70	2	769
OAPEC (b)	2	85	65	–	2	4	158
OLADE (c)	20	265	108	4	33	73	503
Other developing countries	145	368	52	16	27	420	1028
USSR	375	510	535	45	25	–	1490
China	560	90	15	1	7	70	743
Other countries with centrally planned economies	408	117	116	17	4	18	680
World (*)	2692	3143	1723	482	218	601	8859

Notes: (1) Industrialized countries: new and renewable energy sources
           Developing countries: non-commercial energy sources
       (2) Inland consumption + bunkers
       (a) European Economic Community (EUR-10)
       (b) Organization of Arab Petroleum Exporting Countries
       (c) Latin-American Energy Organization.
Methodology: The methodology adopted for the establishment of energy balances is that
              used by the Statistical Office of the European Communities.
       (*) The discrepancy between world production and world consumption is explained
           by possible stock movements and statistical deviation.

TABLE 5 WORLD ENERGY BALANCE, LONG-TERM PROJECTION

2000

A. PRIMARY ENERGY PRODUCTION   million toe

	Solid fuels	Oil	Natur. gas	Nuclear energy	Hydro-electri	Other (1)	Total
EEC(a)	172	108	108	215	14	7	625
USA	939	415	387	180	35	25	1981
Japan	10	2	–	90	11	44	157
Other industrialized countries	457	146	156	104	88	7	958
OAPEC (b)	–	1000	151	..	3	11	1165
OLADE (c)	21	330	139	11	45	85	631
Other developing countries	137	595	141	27	40	525	1465
USSR	420	650	930	105	30	–	2135
China	800	150	15	2	10	90	1067
Other countries with centrally planned economies	500	25	60	40	6	20	651
World (*)	3456	3421	2087	774	282	814	10835

B. GROSS PRIMARY ENERGY CONSUMPTION (2)

Million toe

	Solid fueld	Oil	Natur. gas	Nuclear energy	Hydro-electri	Other (1)	Total
EEC(a)	264	439	196	215	14	7	1136
USA	838	680	455	180	35	25	2213
Japan	109	233	57	90	11	44	544
Other industrialized countries	319	325	139	104	88	7	982
OAPEC (b)	5	104	122	..	3	11	245
OLADE (c)	25	321	138	11	45	85	625
Other developing countries	208	494	69	27	40	525	1363
USSR	420	605	715	105	30	–	1875
China	775	90	15	2	10	90	982
Other countries with centrally planned economies	490	145	157	40	6	20	858
World (*)	3453	3436	2063	774	282	814	10822

Notes: (1) Industrialized countries: new and renewable energy sources
           Developing countries: non-commercial energy sources
       (2) Inland consumption + bunkers
       (a) European Economic Community (EUR-10)
       (b) Organization of Arab Petroleum Exporting Countries
       (c) Latin-American Energy Organization.
Methodology: The methodology adopted for the establishment of energy balances is that
              used by the Statistical Office of the European Communities.
       (*) The discrepancy between world production and world consumption is explained
           by possible stock movements and statistical deviation.

CHAPTER VI

SOME ALTERNATIVES TO THE REFERENCE PROJECTION

A. INTRODUCTION

1. The reference projection described in the previous chapters should
    not be regarded as a central or single forecast but rather as one
    plausible and coherent **scenario** among others.

    The results obtained obviously reflect the assumptions on which the
    projection is based with all their combined uncertainties. Changing
    one or another key assumption would alter the results accordingly.

2. There are two main ways of assessing the influence of the
    assumptions on the results obtained.

    The first is to study alternative and contrasting scenarios which
    together cover a wide range of possible futures. This means
    selecting sets of assumptions which are strictly consistent with
    the socio-economic structures of each of the scenarios studied.
    The study entitled "Scenarios to the year 2000" followed this
    approach.

    The second method is to make a central projection based on a single
    set of consistent assumptions regarded as possible or even
    probable.   The sensitivity of the results to each of the
    assumptions adopted can then be studied by analysing variants on
    each of the main assumptions and their effects on the overall
    results. This second method is adopted for this report.

3. Each of the two methods has specific advantages. They are not
    mutually exclusive but complementary. The first makes it possible
    to examine a wide range of possible futures. The results obtained
    can serve as the background to a study of one or more intermediate
    scenarios that are judged to be "more" probable. The second
    identifies the spread of uncertainty which should be taken into
    account when deciding measures or policies.

4. This chapter briefly describe the results for the Community of the
    variants to the reference projections that have been studied. The
    assumptions that have the greatest impact on the energy market are
    all the subject of such analysis. These concern the rate of
    economic growth, changes in industrial structure over time,
    different import prices for crude oil and for imported natural gas,
    the role of nuclear power, stronger environmental protection
    standards, and finally energy efficiency.

B. ECONOMIC VARIANTS

a) Changes in the rate of economic growth

5. The reference projection postulates an average annual rate of
    economic growth for the Community of 2.6% between now and 2000.

    The first variant tests the impact on the Community's energy
    consumption of a higher rate of growth, 3.5% p.a. between 1985 and
    2000. This variant was worked out **all other things being equal,**
    including the structure of the economy and energy prices. It should
    be pointed out, however, that a higher rate of economic growth
    would encourage increased investment in more energy-thrifty
    processes, thereby improving further the efficiency of the energy
    system.

6. With a GDP 14% higher in 2000 than in the reference projection, the
    Community's gross inland energy consumption would be up by 5.3%. In
    absolute terms, the difference would be 59 million toe, which would
    bring consumption to 1 166 million toe as against 1 107 million toe
    in the reference projection. The reference GDP growth rate of 2.6%
    over the period 1983-2000 would generate an increase in gross
    inland consumption of 1.3% p.a. The economic growth differential of
    0.9% p.a. compared with the reference projection would lead to an
    additional increase in energy consumption of about 0.3% p.a. This
    more marked decoupling of energy demand from GDP would stem from
    the faster replacement of energy production and consumption
    equipment.

7. The increase in gross inland consumption would not bear equally on
    all forms of energy. Most of the new requirements, about 70%,
    would be covered by hydrocarbons. Petroleum products and crude oil
    would supply about 24 million toe and natural gas 16 million toe.
    The consumption increases would mainly occur in the industrial
    sectors and transport whose demand levels are highly sensitive to
    faster GDP growth. As the reference projection already allows a
    fairly large share for solid fuels, consumption of these would
    increase by only 7 million toe, of which a third in industry and
    the remainder in the electricity sector. Finally, in view of the
    lead time for building power stations, faster economic growth would
    not have any significant impact on nuclear power until after 1995.
    By 2000 there would be a 6% increase in demand for nuclear energy,
    i.e. 12 million toe.

8. The larger share of hydrocarbons in gross inland consumption would
    mean increasing the Community's energy imports, because the inland
    production of hydrocarbons adopted in the reference projection is
    already judged to be at the highest feasible level. The increased
    consumption would therefore be covered by imports. In 2000 73% of
    the energy needed to meet additional demand would be imported and
    27%, mainly nuclear, would be produced in the Community.

    Dependence on energy imports would be greater as a result of
    hydrocarbon imports, rising from 45.0% in 2000 in the reference
    projection to 46.4% in this higher growth alternative.

9. A lower rate of growth would have the effect of cutting back gross fixed capital formation and thereby significantly slowing down the replacement of plant and equipment, not least for energy production and consumption. This would slow down the penetration of new energy conversion technologies in the energy system, thereby undermining the medium- and long-term improvement in the European economy's energy efficiency.

    At the same time, however, a slowdown in economic activity and therefore in industrial output would also mean a fall in final demand (half of which is in the industrial sector) and would significantly cut electricity consumption.

10. For example, an annual economic growth rate of only 1.8% between now and 2000 could cut primary energy consumption by nearly 50 million toe compared with the reference projection.

    The drop in electricity consumption would have important implications for the consumption of solid fuels (-10 million toe) and to a lesser extent electronuclear generation (-7 million toe). There could be additional reductions in final consumption, mainly of oil products (-17 million toe) and natural gas (-10 million toe), but also of electricity (-5 million toe) and of solid fuels (-5 million toe).

Summary table: "Economic growth" variant

EUR-10: 2000	Reference projection	"Economic growth" variants	
		Faster GDP growth	Slower GDP growth
Annual GDP growth rate	+2.6%	+3.5%	1.8%
Gross energy consumption (M toe) of which:	1136[1]	1195	1084
Solid fuels	264	271	249
Oil	439	463	421
Natural gas	196	212	186
Nuclear energy	215	227	208
Net oil imports (M toe)	330	354	312
Dependence on imported energy	45.0%	46.4%	43.5%

---
[1] Including bunkers.

Community energy balances: "GDP growth" variants

(a) High economic growth (3.5%)

ENERGY SUPPLY/DEMAND BALANCE			GDP +3.5%		EUR-10		2000 million toe	
	SOLID FUELS	PETROL PRODUCTS	GAS	NUCLEAR ENERGY	HEAT	HYDRO+ OTHERS	RENEW ENERGY	TOTAL
PRIM PRODUCT	172	108	110	227	–	14	8	639
TOT IMPORTS	111	478	118	–	–	4	–	711
TOT EXPORTS	12	124	16	–	–	4	–	156
STOCK CHANGE	–	–	–	–	–	–	–	–
GROSS CONSUMP	271	463	212	227	–	14	8	1195
BUNKERS	–	29	–	–	–	–	–	29
INLAND CONSUMP	271	434	212	227	–	14	8	1166
ELEC POWER STNS	181	20	18	227	-15	-155	4	280
OTHER TRANSF	24	4	-15	–	–	–	–	14
ENERG INDUST	2	30	11	–	–	25	–	68
FINAL CONSUMP	64	380	198	–	15	144	4	804
– NON ENERGY	2	60	13	–	–	–	–	75
– INDUSTRY	48	59	82	–	5	61	–	255
– TRANSPORT	–	187	0	–	–	4	–	191
– HOUSEHOLD	14	74	104	–	10	80	4	286

(b) Low economic growth, (1.8%)

ENERGY SUPPLY/DEMAND BALANCE			GDP +1.8%		EUR-10		2000 million toe	
	SOLID FUELS	PETROL PRODUCTS	GAS	NUCLEAR ENERGY	HEAT	HYDRO+ OTHERS	RENEW ENERGY	TOTAL
PRIM PRODUCT	172	108	104	208	–	14	6	612
TOT IMPORTS	89	436	98	–	–	4	–	627
TOT EXPORTS	12	124	16	–	–	4	–	156
STOCK CHANGE	–	–	–	–	–	–	–	–
GROSS CONSUMP	249	421	186	208	–	14	6	1084
BUNKERS	–	29	–	–	–	–	–	29
INLAND CONSUMP	249	392	186	208	–	14	6	1055
ELEC POWER STNS	168	18	15	208	-15	-142	4	256
OTHER TRANSF	23	4	-15	–	–	–	–	13
ENERG INDUST	2	28	11	–	–	23	–	63
FINAL CONSUMP	56	342	175	–	15	133	2	723
– NON ENERGY	2	56	12	–	–	–	–	70
– INDUSTRY	42	47	67	–	5	55	–	216
– TRANSPORT	–	169	0	–	–	4	–	173
– HOUSEHOLD	12	70	96	–	10	74	2	264

b) Changes in the structure of the economy

11.   The reference projection is based on the assumption that services and the manufacturing and capital goods industries will increase their shares in the structure of GDP at the expense mainly of the energy-intensive industries.   This assumption has a favourable impact on the primary energy intensity of the Community economy.

How important is that impact? To assess this we retained the same GDP growth rate as in the reference projection for the whole period, but we assumed that the economic structure would not undergo further change as from 1985. At the same time, however, we assumed that the unit energy consumption cuts witnessed in the principal consumer sectors in the reference scenario were maintained. The variant therefore provides an indication purely of the "structural effects".

12. The increase in final energy consumption resulting from this alternative is about 1.7% in 1990 (12 million toe), rising to 5.5% in 2000 (42 million toe). Final energy consumption per unit of GDP falls more slowly in this variant than in the reference projection, from 0.49 toe/'000 ECU in 1983 to 0.40 in 2000 as against 0.384 in the reference projection. This is the result of the larger share of the energy-intensive industries (steel, basic chemicals, non-ferrous metals, etc.) in the structure of GDP.

In 2000, taking account of a significant increase in electricity demand by industrial sectors (+8 million toe), gross inland primary energy consumption resulting from an economic structure fixed in its present shape would be 60 million toe higher than in the reference projection, an increase of the order of 5.3%.

13. This increase is of the same order of magnitude as that obtained in variant (a) (impact of a GDP growth rate of 3.5% instead of 2.6%). In exactly the same way, in this case the only element in the Community's energy production to increase would be nuclear power, additional fossil fuels being imported. The rate of dependence on imports would increase from 45.0% in 2000 in the reference projection to 46.4% in this variant. But the breakdown of gross inland energy consumption by product reveals much sharper differences between the two variants.

14. Solid fuels show an increase of 18 million toe in 2000 compared with the reference projection. This figure is higher than that obtained in the GDP growth variant. This is because in this case steel and cement works, both large consumers of coke and coal, grow at the same average annual rate as GDP between 1985 and 2000, i.e. 2.6% (keeping their share in the economic structure), whereas in the first variant these sectors grow more slowly as a result of the shift to services and capital goods in the structure of economic activity. The consumption of solid fuels by the industrial sectors alone would increase by nearly 11 million toe in 2000, while the electricity sector would also increase its offtake as a result of the higher electricity consumption by industrial sectors.

In 2000, an additional 18 million toe of crude oil and petroleum products would also be necessary to meet the increase in demand resulting from the 1985 economic structure. These fuels are consumed mainly by industry (+12 million toe), mostly in the form of heavy fuel oil but also as petrochemical feedstocks. The induced demand for motor fuel for goods transport also rises (+2 million toe) as a result of increased industrial activity. Finally there would be a slight increase in consumption by the electricity sector (+3 million toe). More than half the increase in consumption of oil products would therefore be met by heavy or extra heavy fuel oil, which would somewhat tilt the structure of the barrel of consumption towards heavier products. However, the changes would not be enough to call in question efforts to rationalize the oil sector.

As to final energy consumption, gas deliveries would increase by 12 million toe in 2000, a similar figure to that for primary consumption in which additional natural gas requirements are put at 13 million toe. This equilibrium between final demand and primary requirements is the net result of two opposing movements: an increase in the consumption of natural gas by power stations (+2 million toe); and a greater output and therefore use of coking and blast furnace gas due to the increase in steel output (compared with the reference projection).

Finally, nuclear energy production shows no change from the reference projection up to 1993/94 because of the time required to build power stations, but will rise substantially thereafter to meet the increase in electricity consumption in industry, its regular consumers. In 2000 the production of nuclear energy could increase by 13 million toe, the equivalent of 50 TWh. Expressed in capacity terms, this corresponds approximately to the installation of an additional 8 GW.

Summary table: "Economic structure" variant

EUR-10: 2000	Reference projection	"Economic structure" variant
Annual GDP growth rate	+2.6%	2.6%
Gross energy consumption (M toe)	1136	1196
of which:		
solid fuels	264	282
oil	439	457
natural gas	196	209
nuclear energy	215	228
Net oil imports (M toe)	330	348
Dependence on imported energy	45.0%	46.4%

Community energy balance: "Economic structure" variant

ENERGY SUPPLY/DEMAND BALANCE   EUR-10   2000
                                                                          million toe

	SOLID FUELS	PETROL PRODUCTS	GAS	NUCLEAR ENERGY	HEAT	HYDRO+ OTHERS	RENEW ENERGY	TOTAL
PRIM PRODUCT	172	108	112	228	–	14	7	641
TOT IMPORTS	122	472	113	–	–	4	–	711
TOT EXPORTS	12	124	16	–	–	4	–	156
STOCK CHANGE	–	–	–	–	–	–	–	–
GROSS CONSUMP	282	457	209	228	–	14	7	1196
BUNKERS	–	29	–	–	–	–	–	29
INLAND CONSUMP	282	428	209	228	–	14	7	1167
ELEC POWER STNS	182	21	17	228	–15	–156	4	281
OTHER TRANSF	26	4	–17	–	–	–	–	13
ENERG INDUST	2	30	12	–	–	25	–	68
FINAL CONSUMP	72	373	197	–	15	145	3	805
– NON ENERGY	2	61	14	–	–	–	–	77
– INDUSTRY	57	60	84	–	6	66	–	273
– TRANSPORT	–	180	0	–	–	4	–	184
– HOUSEHOLD	13	72	99	–	9	76	3	271

C. ENERGY PRICE VARIANTS

a) Crude oil import prices

15.   The central assumption in the Community reference projection for
the average cost of imported crude oil was $35/bbl in 2000 (at 1983
prices), after a decline in the nominal price to $27/bbl in 1985
and holding firm at this level in real terms until 1990.

Several conditions would have to be met for this assumption to be
borne out, of which the most important is that there will be no
major short or medium-term structural imbalances between demand and
supply on the international oil market. Such an imbalance currently
exists and has driven prices down much more sharply in the short
term than was assumed for the reference scenario.

16.   Against that background two alternative price scenarios were
studied in the sensitivity analyses.

The first variant takes the assumption of an oil price eroding
steadily in nominal terms up to 1990 and then remaining stable in
real terms at $20/bbl up to 2000.

The second variant takes the contrary assumption in which, for
whatever reason, under pressure of a relative scarcity of world
supply after 1985, the oil price rises rapidly in real terms to
$40/bbl (at 1983 prices) in 1990 and $50/bbl in 2000.

For each of these variants two essential elements are considered:

- the general effect on the Community economy and the induced effect on total energy demand;

- the impact on energy sources competing with oil and the induced substitutions between different forms of energy.

17.   As regards economic activity, the "low price" variant crude oil price would give, all other things being equal, a faster increase in private consumption and investment than in the reference projection, mainly as a result of the less acute pressure on the Community's balance-of-payments.

This would bring about a higher annual rate of change in GDP than in the reference projection, at +3% on average between 1990 and 2000. The associated growth rate of gross internal energy consumption would be 1.8% p.a. over the same period. Over the period 1990-2000 the ratio of the growth rate of energy demand to that of GDP would be 0.60 as against 0.37 in the reference projection. This increase in the energy intensity of the European economy would stem from the significant lengthening of the assured payback periods for energy savings and oil-substitution investments in all economic sectors. These longer payback times would undermine the economic profitability of many such projects and lead to some abandonment. In 2000 the Community's gross energy consumption including bunkers would reach 1 218 million toe, which is 82 million toe more than in the reference projection.

18.   Lower crude oil import prices would have important implications for competing energy sources. The Community's internal sources of hydrocarbon supply would remain largely competitive up to 2000 despite the abandoning of certain new planned production investment, the profitability of which would no longer be assured. It can also be assumed that natural gas imports would increase somewhat for two reasons: in the reference projection they are close to the **minimum** offtake thresholds under present or planned contracts, and with gas prices indexed in one way or another to petroleum product prices, gas would also benefit in competitiveness from the drop in the crude oil price.

In contrast, the competitiveness of imported coal or of home-produced lignite and peat could be put at risk in 2000, increasing the pressures on the Community's coal output. Community production might have to decline by 20 million to 30 million toe, while Community aids to remaining production capacities would increase by about 25%.

19.   For the total energy balance, the low oil price variant would trigger three movements:

- contracting demand for solid fuels, which would increase by only 8.0% between 1983 and 2000 instead of the 25% shown in the reference projection. This trend would stem mainly from the lower competitiveness of electricity compared with oil products and natural gas for space heating, and from a slower rate of introduction of new coal-fired power stations – both of which would reduce power station demand for coal;

– a fairly sharp recovery in oil demand as a result both of rising energy demand and the shift away from the direct and indirect (via electricity) use of coal;

– a concomitant increase in demand for natural gas. Through the indexation systems it would maintain its market shares among the main energy consumers in an expanding market. The increase in Community consumption of natural gas would however be constrained by the likely reduction in volumes of gas potentially available on the international natural gas market at such low prices. The substantially lower contract prices for natural gas would affect the profitability of producing countries which are either operating in difficult conditions (offshore fields) or use the LNG process for transmission.

20. The second imported crude oil price variant ($50/bbl in 2000 instead of $35/bbl) takes an oil price that is rising rapidly after 1985 (+10% p.a. up to 1990) under the pressure, for example, of a destabilized oil market. After 1990, the annual rate of increase in the import price would lose some momentum.

In this assumption, the real rate of Community GDP growth would be lower than that shown in the reference projection, mainly as a result of the inflationary effect of higher energy prices and the additional balance-of-payments cost of oil imports. The average annual GDP growth rate over the period 1985-90 would barely reach 2% although between 1990 and 2000 it would probably be closer to 2.5%, making an average annual GDP growth rate of 2.3% over the period 1983-2000.

Gross inland energy consumption would increase at an average annual rate of 0.9% over the same period, amounting (including bunkers) to 1066 million toe in 2000, i.e. 70 million toe or 6% less than in the reference projection. Between 1983 and 2000 the ratio of energy demand growth to GDP growth would thus be 0.4 as against 0.52 in the reference projection. A sharp rise in the price of energy would substantially increase the profitability of energy savings investments, reducing their payback time and thus promoting their adoption by major energy consumers despite the slower economic growth which the price rise would bring about.

21. At an average imported crude oil price of $50/bbl, natural gas of whatever origin would break the threshold of competitiveness. Using the same price formation mechanisms as in the reference projection, the average procurement cost of natural gas in the Community would be $38.5/bbl or $6.2/MBTU.

At that price, virtually all the fields now known would be economically exploitable. Gas deliveries from different origins would then be in competition against each other in a slow-growth global energy market. Such abundance of supply on the gas market could hold down the selling prices of natural gas to the marginal cost of production, estimated in this variant at around $5.6/MBTU or $35/bbl, including the gas royalties normally demanded by the producing countries. In 2000 even this price level would given natural gas a relative price advantage of nearly 15% on the inland markets compared with oil products. This difference would strengthen the shift from oil products to natural gas for all non-specific uses such as heating, steam-raising or petrochemical feedstock. Only the rate of equipment replacement and the need to amortize previous investments as far as possible in the context of slower economic growth could temper this trend.

22. Besides natural gas electricity should also become more competitive, mainly in thermal uses, which would increase its share in meeting final energy requirements from 18.1% in the reference projection to 21.7% in this variant. This increase in electricity consumption of about 125 TWh would of course be reflected in the consumption of fuel for electricity generation, mainly nuclear fuel and solid fuels.

The time needed for building nuclear power stations in the Community now averages nearly 10 years. Any increase in nuclear capacity to be available in 2000 must therefore be planned

before 1990. Given also the lapse of two to three years which
investors need to react to the price rises expected in this
scenario after 1985, it may be assumed that the additional nuclear
capacity that can be brought on stream in 2000 would hardly exceed
an additional 5 GW per year, which would be concentrated in the
last three or four years of the 1990s. Thus additional capacity of
about 15 GWe would cover 75 TWh of increased demand. The shortfall
would be met by solid fuels, which are better suited than nuclear
fuels for adjusting to the load curves resulting from the thermal
uses of electricity.

23.   The Community energy balance resulting from the high oil price
assumption would exhibit the following features:

- a sharp reduction in the rôle of oil, which in 2000 would cover
  only a quarter of energy requirements, the same share as other
  forms of energy;

- a major switch to natural gas in the final consumption sectors
  where it would compete with oil products in all their specific
  applications;

- increased use of electricity for thermal uses, and a concomitant
  further development of nuclear energy.

Summary table: oil price variants

EUR-10 2000	Reference projection	"Oil price" variants	
		low price	high price
Average import price of crude oil	35$/bbl	20$/bbl	50$/bbl
Annual GDP growth rate	+2.6%	2.8%	+2.3%
Gross energy consumption (million toe)	1136	1218	1066
of which:			
Solid fuels	264	229	275
Oil	439	539	306
Natural gas	196	216	226
Nuclear energy	215	212	230
Net oil imports (million toe)	330	439	194
Dependence on imported energy	45.0%	50.9%	38.7%

Community energy balances: oil price variants

(a) Variant   20$/bbl

ENERGY SUPPLY/DEMAND BALANCE   EUR-10   2000
Million toe

	SOLID FUELS	PETROL PRODUCTS	GAS	NUCLEAR ENERGY	HEAT	HYDRO+ OTHERS	RENEW ENERGY	TOTAL
PRIM PRODUCT	152	100	112	212	-	14	7	597
TOT IMPORTS	89	563	120	-	-	4	-	776
TOT EXPORTS	12	124	16	-	-	4	-	156
STOCK CHANGE	-	-	-	-	-	-	-	-
GROSS CONSUMP	229	539	216	212	-	14	7	1218
BUNKERS	-	29	-	-	-	-	-	29
INLAND CONSUMP	229	510	216	212	-	14	7	1189
ELEC POWER STNS	156	22	19	212	-13	140	4	260
OTHER TRANSF	21	4	-15	-	-	-	-	10
ENERG INDUST	2	36	12	-	-	23	-	73
FINAL CONSUMP	50	448	200	-	13	131	3	846
- NON ENERGY	2	60	13	-	-	-	-	75
- INDUSTRY	38	82	79	-	5	54	-	258
- TRANSPORT	-	197	0	-	-	5	-	202
- HOUSEHOLD	10	109	108	-	8	73	3	311

(b) Variant   50$/bbl

ENERGY SUPPLY/DEMAND BALANCE   EUR-10   2000
Million toe

	SOLID FUELS	PETROL PRODUCTS	GAS	NUCLEAR ENERGY	HEAT	HYDRO+ OTHERS	RENEW ENERGY	TOTAL
PRIM PRODUCT	172	112	110	230	-	16	13	653
TOT IMPORTS	115	318	132	-	-	4	-	569
TOT EXPORTS	12	124	16	-	-	4	-	156
STOCK CHANGE	-	-	-	-	-	-	-	-
GROSS CONSUMP	275	306	226	230	-	16	13	1066
BUNKERS	-	29	-	-	-	-	-	29
INLAND CONSUMP	275	277	226	230	-	16	13	1037
ELEC POWER STNS	190	12	15	230	-18	157	6	280
OTHER TRANSF	22	3	-15	-	-	-	-	11
ENERG INDUST	2	20	13	-	-	24	-	59
FINAL CONSUMP	61	242	216	-	18	149	7	687
- NON ENERGY	2	50	15	-	-	-	-	67
- INDUSTRY	44	24	86	-	6	61	2	223
- TRANSPORT	-	136	0	-	-	6	-	142
- HOUSEHOLD	15	32	112	-	12	82	5	258

(b) Natural gas prices

24. The reference projection assumes that the "guide price" for energy
    will continue to be that of oil, and that the mechanisms for
    determining gas prices and indexing them on crude oil and/or other
    oil products will continue to apply up to 2000. This would mean a
    natural gas consumption of 196 million toe in 2000, 18% of total
    Community inland energy consumption, against 165 million toe and
    19% in 1983.

    This level of consumption is only 20 million toe above supplies
    contracted for by end 1984. Since this is obviously well below the
    potential supply available from the international market, it is
    possible to imagine that entirely different mechanisms for the
    formation of natural gas prices could be introduced in the medium
    term.

    In order to secure its place in the energy system, in a scenario of
    moderate growth of energy consumption natural gas must be able to
    remain competitive against the energies it could supplant: coal and
    heavy fuel oil on the industrial markets, heating oil on the
    residential and tertiary markets. Substitution for coal and heavy
    fuel oil will be the determining factor, because the gas market
    needs industrial customers if it is to expand. Industry is a major
    potential market where the demand is fairly stable and can be
    modulated, unlike residential uses with their sudden peaks and
    troughs.

25. Even under the reference oil price scenario the formation and
    indexing of natural gas prices could change, with prices no longer
    primarily based on the costs of oil supplies but rather on a
    mixture of the consumer prices of coal, heavy fuel oil and heating
    oil. Such a formula could bring down the price of natural gas by
    about 6-10%, compared with the reference scenario, depending on the
    weighting indices adopted.

    Under this assumption, what would be the largest share which
    natural gas could hold on the energy market in 2000?

    On the industrial market, gas could increase its market by about
    9 million toe, mainly by replacing heavy fuel oil, but also coal,
    to a lesser extent. Together with its comparative price advantage,
    natural gas also offers extreme ease of exploitation and the virtue
    - important for the future - of being practically non-polluting.

    For non-energy uses, the synergetic relationship between the large
    petrochemical industries and the oil companies would make major
    trade-offs between naphta and natural gas improbable, given the
    price differential discussed here.

    In the residential and tertiary sectors, consumption could increase
    if there was a faster rate of connection to the network in areas
    already supplied with natural gas. But the lower price would not
    lead to any significant expansion of distribution networks in areas

not yet supplied. Accordingly the increase in consumption in this sector could be around 8 million toe, almost all of it displacing heating oil.

Finally, sales to power stations would not rise significantly if the increases in consumption were equally shared between the residential and industrial sectors.

26. Overall, a drop in natural gas consumer prices of 6-10% would allow the market to absorb an additional 20 million toe of natural gas, an increase of 10%. Since the best international prices at the moment are those of the USSR and the Netherlands, these countries would be best placed to meet this increase in consumption, especially as the transport facilities between them and the Member States are already available. (The trans-European gas pipeline between the USSR and Western Europe is at present under-utilized).

Summary table: Reference projection;
"natural gas price" variant

EUR-10: 2000	Reference projection	"Natural gas price" variant
Average procurement price of natural gas	4.34$/MBTU	4.0$/MBTU
Gross energy consumption (mtoe) of which:	1136	1136
Solid fuels	264	259
Oil	439	424
Natural gas	196	216
Nuclear energy	215	215
Net oil imports (mtoe)	330	315
Dependence on imported energy	45.0%	45.0%

27. As shown above (paragraphs 20 and 21), a fast rise in the oil price under the pressure of rapidly increasing world oil consumption could upset the basic parameters of the gas market. Assuming a price per barrel of imported oil of $50 in 2000 (at 1983 prices) and on the basis of present natural gas indexing mechanisms, the selling price of natural gas in 2000 would average $6.20/MBTU ($38.5/bbl) as against $3.80/MBTU in early 1984. In this case all known potential sources of natural gas (North Sea, northern Norwegian, Africa, etc.) would become competitive. The additional supply of natural gas on the European market could reach nearly 100 million toe in 2000.   In such a market situation supply pressures could coceivably free gas more completely from its link with oil, and its price should theoretically level off at the

highest marginal cost of development. Gas supplies from various possible origins would then compete against each other. In this case as we have seen, Community consumption of natural gas could amount to 226 million toe in 2000 and thus absorb 50% of the potential additional supply of natural gas in that year.

In a situation of a natural gas bubble on the Community market at the prices suggested, we could witness an even greater fall in prices similar to that on the oil market from 1982 to 1985. In an extreme scenario, such a downward spiral of prices could bring about a drop of the order of 30% relative to those resulting from present indexing mechanisms. This presupposes of course that the exporting countries with the highest production costs give up part of their gas royalties in order to acquire new market shares. It is important to note that for the gas exporting countries what is principally at stake is the period after 2000: they need to secure the best possible position as soon as they can with regard to their potential purchasers.

28. Compared with energy consumption in the high crude oil price variant ($50/bbl), additional deliveries of natural gas under this scenario of extreme competition price scenario could approach 30 million toe, assuming that the prices of other primary energy sources did not tend to align on the new natural gas prices. Imports from outside the Community would then remain within the limits of supply generally regarded as accessible.

As regards the consumer sectors, natural gas would be able to compete with the oil products used as feedstocks in the basic chemicals industry and would put up a stronger resistence to any marked expansion of electricity in covering thermal uses. The gap between the price of natural gas and that of motor fuels would warrant greater use of natural gas for vehicles, as already happens on some local markets. Natural gas could also marginally replace solid fuels and oil products for electricity generation. It should be noted here that falling demand for electricity would slow down the development of additional electronuclear capacities.

Summary table: Scenario $50/bbl;
natural gas prices variant

EUR-10: 2000	Reference projection	Scenario $50/bbl	
		High gas price	Low gas price
Average price of imported crude oil	35$/bbl	50$/bbl	50$/bbl
Average procurement price of natural gas	4.3$/MBTU	5.6$/MBTU	4.4$/MBTU
Gross energy consumption (mtoe)	1136	1066	1068
of which:			
Solid fuels	264	275	270
Oil	439	306	290
Natural gas	196	226	256
Nuclear energy	215	230	222
Net oil imports (mtoe)	330	194	178
Dependence on imported energy	45.0%	38.7%	39.6%

Overall Community balances, "natural gas prices" variants

(a) Reference projection; revised indexing mechanisms

ENERGY SUPPLY/DEMAND BALANCE   EUR-10   2000

million toe

	SOLID FUELS	PETROL PRODUCTS	GAS	NUCLEAR ENERGY	HEAT	HYDRO+ OTHERS	RENEW ENERGY	TOTAL
PRIM PRODUCT	172	108	108	215	–	14	7	625
TOT IMPORTS	99	439	124	–	–	4	–	667
TOT EXPORTS	12	124	16	–	–	4	–	156
STOCK CHANGE	–	–	–	–	–.	–	–	–
GROSS CONSUMP	259	424	216	215	–	14	7	1136
BUNKERS	–	29	–	–	–	–	–	29
INLAND CONSUMP	259	395	216	215	–	14	7	1107
ELEC POWER STNS	176	17	18	215	–15	–148	4	266
OTHER TRANSF	23	4	–15	–	–	–	–	13
ENERG INDUST	2	29	11	–	–	24	–	65
FINAL CONSUMP	58	345	202	–	15	138	3	763
– NON ENERGY	2	58	12	–	–	–	–	72
– INDUSTRY	43	45	82	–	5	58	–	234
– TRANSPORT	–	178	0	–	–	4	–	182
– HOUSEHOLD	13	64	108	–	10	77	3	274

(b) Variant $50/bbl; 30% reduction in the price of natural gas

ENERGY SUPPLY/DEMAND BALANCE   EUR-10   2000

million toe

	SOLID FUELS	PETROL PRODUCTS	GAS	NUCLEAR ENERGY	HEAT	HYDRO+ OTHERS	RENEW ENERGY	TOTAL
PRIM PRODUCT	172	112	110	222	–	16	13	645
TOT IMPORTS	110	302	162	–	–	4	–	578
TOT EXPORTS	12	124	16	–	–	4	–	156
STOCK CHANGE	–	–	–	–	–	–	–	–
GROSS CONSUMP	270	290	256	222	–	16	13	1068
BUNKERS	–	29	–	–	–	–	–	29
INLAND CONSUMP	270	261	256	222	–	16	13	1039
ELEC POWER STNS	185	10	22	222	–18	154	6	273
OTHER TRANSF	22	3	–15	–	–	–	–	11
ENERG INDUST	2	18	15	–	–	24	–	59
FINAL CONSUMP	61	230	234	–	18	146	7	696
– NON ENERGY	2	43	23	–	–	–	–	68
– INDUSTRY	44	24	90	–	6	60	2	226
– TRANSPORT	–	131	5	–	–	6	–	142
– HOUSEHOLD	15	32	116	–	12	80	5	260

D.   ALTERNATIVES RELATING TO DEMAND FOR SOLID FUELS

29.   Despite an import price of non-Community coal assumed to be based
      on the cost of developing new mines in the main exporting countries
      (which tends to reduce the differential between the price of coal
      and the price of oil), there will nevertheless be a major increase
      in coal consumption in the Community between 1983 and 2000 under
      the reference scenario. Coal consumption will expand mainly in the
      electricity sector (+33%), gradually displacing hydrocarbons in
      present power stations and it should also make something of a
      comeback in industry and for heating purposes.

      Quite apart from any consideration of its international price, two
      factors could change the position of coal on the European energy
      scene:

      – a slowdown in the expansion of nuclear energy if there should be
        another major incident which has the impact of Three Mile Island;

      – the introduction of stricter standards for SO_x and NO_x emissions
        for all large combustion plants.

30.   As far as a nuclear slowdown is concerned we took the assumption
      that no new stations would be commissioned. Any nuclear moratorium
      which put a stop to the building of new power stations as early as
      1985 would of course have a major impact on the functioning of the
      electricity sector, as nuclear energy is one of the main targets
      for capacity expansion in most Member States. Outside the
      electricity sector, it would have significant repercussions for the
      general equilibrium of the energy system.

The assumption is that the nuclear power stations not built would be replaced mainly by coal stations, coal being the cheapest fossil energy source under the price scenario adopted. But natural gas and, more marginally, oil products could also benefit as better use is made of available capacities. If environmental protection measures are made stricter, the use of natural gas could rise considerably since it is the only form of energy whose emission levels, at virtually nil, can be compared to those of nuclear energy.

The switch of investment from nuclear to conventional power stations would increase the production cost of electricity and make it less competitive, especially for heating uses. This could lead to limited substitution by natural gas and, to a lesser extent, by oil products.

Leaving aside these particular considerations, the share of nuclear energy in electricity generation in 2000 would be reduced to 30.3% against 42.9% in the reference projection. The shift would be to solid fuels, whose share would rise from 39.7% to 50.4%, and hydrocarbons (from 6.9% to 8.7%).

In the primary energy balance, the fall in consumption of nuclear fuel (-65 million toe) would be offset by an increase in coal consumption (+45 million toe), natural gas (+12 million toe) and oil products (+5 million toe). In either case, the increase in consumption could be met only by additional imports. This should be no problem for gas and oil products, but for coal there could be problems in port infrastructures and inland distribution facilities, not to mention the pressure such substantial additional demand would put on international prices.

Summary table: "nuclear moratorium" variant

EUR-10: 2000	Reference projection	"Nuclear moratorium" variant
Installed nuclear capacity	144 GWe	100 GWe
Gross energy consumption (mtoe)	1136	1133
of which:		
Solid fuels	264	309
Oil	439	444
Natural gas	196	208
Nuclear energy	215	150
Net oil imports (mtoe)	330	335
Dependence on imported energy	45.0%	50.4%

31. With its proposal for a Directive on the limitation of emissions of pollutants into the air from large combustion plants,[2] the Commission is seeking to take action against air pollution in the form of acid rain and the damage it causes to forests and to lakes, farmland, monuments and buildings. The aim of the Directive is to reduce SO_2 emissions by 60% and NO_x emissions by 40% in each Member State by 1995, the reference year being 1980. All combustion plants with a nominal thermal output greater than 50 MW for which a construction permit is issued after 1 January 1985 would be required to comply with the emission values set by the Directive. These are shown in the following table.

-- -------- --------------
[2] COM(83)704.

Emission limit values for sulphur dioxide and oxides of nitrogen
(in mg/m³ of waste gases under standard conditions)

Pollutant / e of fuel	Sulphur dioxide[1]		Oxides of nitrogen (as NO_2)	
	as from 1/1/1985	after 31/12/1995	as from 1/1/1985	after 31/12/1985
id	≤400 as a rule (≤2000 for grate firing or pulverized coal firing where the output is less than 300 MW).	≤250 as a rule	≤800 as a rule (≤1300 for pulverized coal firing with extraction of fused ash).	≤400 as a rule (≤800 for pulverized coal firing with extraction of fused ash).
uid	≤400 as a rule (≤1700 for plants whose output is less than 300 MW).	≤250 as a rule	≤ 450	≤220
eous	≤35 as a rule (≤100 for coke oven gas) (≤5 for liquefied gas)		≤ 350	≤180

count should be taken of the proportion of sulphur trioxide in the waste gas.

32.   This Directive would certainly have repercussions for the structure
      of energy consumption by industry and the electricity sector.

      For electricity, an analysis of the implications by Member State
      shows that:

      - countries launching a major nuclear energy development programme
        will not be much affected by the introduction of the Directive,
        since the addition of nuclear power stations to electricity
        generating capacity will drastically reduce SO_2 and NO_x
        emissions;

      - countries in which liquid fuels still accounted for a large share
        of electricity generation in 1980 and which are reducing
        consumption by replacing heavy fuel oil by natural gas or coal
        will see a sharp reduction in emissions, as fuel oil has a higher
        sulphur content;

      - countries in which coal supplied a large share of electricity
        generation in 1980 and whose nuclear energy programmes, if any,
        will meet only the increase in electricity demand will be obliged

to equip 25% to 55% of their capacity with flue gas desulphuration (FGD) in order to reduce emissions by 60% overall in 1995.

The methods available to industry and electricity producers for reducing their SO_2 and NO_x emissions include:

- replacing the present fuel by a less polluting fuel. Natural gas would be a useful solution as it does not require any additional purifying equipment;

- using fuels with a low sulphur content (imported LSC coal, fuel oil with 1-1.5% sulphur content), which will reduce emissions at the cost of a 4-8% increase in the price of coal and of $35/tonne for residual heavy fuel oil whose sulphur content is reduced from 3% to 1.5%. Using these fuels will not be enough to satisfy the standards in the Directive for new installations;

- installing desulphuring units (FGD), which is a costly investment (from $90/KWe for large output plants to $200/KWe for low output plants). Installing this equpiment is economic only for medium-size plants ($>$200 MWth), which excludes most industrial installations. For these plants it will be necessary to use low-sulphur fuel until new combustion techniques are introduced such as fluidized bed combustion and coal gasification.

Using these methods in order to comply with the stricter emission standards in conventional power stations will increase production costs by 8-12%, depending on the Member State. In this case, total final electricity demand could be cut by 3-4% in 2000. The drop in consumption would mainly be in the industrial sectors, which are the most sensitive to any increase in the price of electricity. Natural gas would be the main substitute fuel, together with oil products to a lesser extent.

33. On the basis of all these considerations a projection can be made
     for changes to the energy balance.   For the electricity sector,
     electronuclear programmes would be speeded up, especially in the
     Member States in which coal now plays a large part in electricity
     generation.   This substitution would be almost entirely at the
     expense of coal and would concern the base load.   In conventional
     power stations with a low rate of utilization ($<$1500 hours), there
     will be a partial substitution of natural gas for coal and fuel oil
     in order to avoid substantial FGD investments.

     For the electricity sector as a whole 28 million toe of coal and
     2 million toe of oil products could be replaced by 12 million toe
     of natural gas and 20 million toe of nuclear fuel; this calculation
     does not take account of effects stemming from any decline in
     electricity generation.

     It is more difficult to offer an estimate for industry as the plant
     and equipment concerned are so diverse.   However, the industrial
     sectors will have to make efforts to reduce emissions (on a scale
     varying from one country to another) in order to reach the required
     overall reduction of 60% (SO_2) laid down in the Directive.   The net
     result of all the substitutions would be a shift to natural gas of
     16 million toe at the expense of coal (3 million toe), oil products
     (5 million toe) and electricity (5 million toe).

     In total inland energy consumption the share of coal would fall by
     41 million toe and that of oil products by 9 million toe, while the
     share of natural gas would increase by 30 million toe and that of
     nuclear fuels by 20 million toe.

Summary table: "Stricter emission standards" variant

EUR-10: 2000	Reference projection	"Stricter emission standards" variant
Gross energy consumption (mtoe) of which:	1136	1136
Solid fuels	264	223
Oil	439	430
Natural gas	196	226
Nuclear energy	215	235
Net oil imports (mtoe)	330	321
Dependence on imported energy	45.0%	43.8%

Overall Community balances: "solid fuel" variants

(a) Nuclear moratorium variant

ENERGY SUPPLY/DEMAND BALANCE   EUR-10   2000
million toe

	SOLID FUELS	PETROL PRODUCTS	GAS	NUCLEAR ENERGY	HEAT	HYDRO+ OTHERS	RENEW ENERGY	TOTAL
PRIM PRODUCT	175	108	108	150	–	14	7	562
TOT IMPORTS	146	459	116	–	–	4	–	726
TOT EXPORTS	12	124	16	–	–	4	–	156
STOCK CHANGE	–	–	–	–	–	–	–	–
GROSS CONSUMP	309	444	208	150	–	14	7	1133
BUNKERS	–	29	–	–	–	–	–	29
INLAND CONSUMP	309	415	208	150	–	14	7	1104
ELEC POWER STNS	223	22	22	150	-15	-146	4	260
OTHER TRANSF	23	4	-15	–	–	–	–	13
ENERG INDUST	2	29	11	–	–	24	–	65
FINAL CONSUMP	61	360	190	–	15	136	3	765
– NON ENERGY	2	58	12	–	–	–	–	72
– INDUSTRY	46	51	75	–	5	57	–	235
– TRANSPORT	–	178	0	–	–	4	–	182
– HOUSEHOLD	13	73	103	–	10	75	3	276

(b) Stricter emission standards variant

ENERGY SUPPLY/DEMAND BALANCE   EUR-10   2000
million toe

	SOLID FUELS	PETROL PRODUCTS	GAS	NUCLEAR ENERGY	HEAT	HYDRO+ OTHERS	RENEW ENERGY	TOTAL
PRIM PRODUCT	165	108	108	235	–	14	7	638
TOT IMPORTS	70	445	134	–	–	4	–	653
TOT EXPORTS	12	124	16	–	–	4	–	156
STOCK CHANGE	–	–	–	–	–	–	–	–
GROSS CONSUMP	223	430	226	235	–	14	7	1136
BUNKERS	–	29	–	–	–	–	–	29
INLAND CONSUMP	223	401	226	235	–	14	7	1107
ELEC POWER STNS	140	16	26	235	-15	-143	4	263
OTHER TRANSF	23	4	-15	–	–	–	–	13
ENERG INDUST	2	27	13	–	–	24	–	65
FINAL CONSUMP	58	354	201	–	15	133	3	766
– NON ENERGY	2	58	12	–	–	–	–	72
– INDUSTRY	43	48	84	–	5	55	–	236
– TRANSPORT	–	178	0	–	–	4	–	182
– HOUSEHOLD	13	70	105	–	10	75	3	275

E. <u>DIFFERENT ASSUMPTIONS ABOUT ENERGY EFFICIENCY</u>

34. The assumptions adopted for the reference projection suggest that the primary energy intensity of the Community economy should decline by about 20% between 1983 and 2000. This figure implies that current energy-saving and RUE policies are continued and takes account of structural changes affecting the components of GDP. The end result in 2000 should be a decline in energy consumption of about 260 million toe, of which about 60 million toe will result from structural changes (shift in the industrial structure to less energy-intensive industries, growing share of services) and 200 million toe obtained from more efficient use of energy.

35. Although this is a substantial improvement, it does not exhaust the possibilities for more efficient energy use in the final consumption sectors. If much more determined and vigorous policies were introduced generally throughout the Community, the additional savings could range - according to different studies - from about 100 to 120 million toe compared with the reference projection. The Community's total primary energy demand in 2000 would be reduced to 1030 million toe, making an average annual growth rate for energy demand of 0.7% over the period 1983-2000. This would reduce the primary energy intensity of the Community economy by a further 7% between 1983 and 2000 and would bring down the elasticity of energy consumption with respect to GDP to 0.28 over the same period.

36. This further reduction in the Community's energy consumption would be the result of intensified energy-saving policies in such areas as:

- **insulation of buildings.** Stricter heat-loss standards would reduce the heating requirements of the residential, commercial and tertiary sectors by up to 15%;

- **heat recovery in industry** by making greater use of heat exchangers, which could reduce up to 6% of total energy consumption by industry;

- the accelerated development of **combined heat and power** production both to meet industrial requirements and to supply heat to district heating systems;

- **heat pumps;**

- **electronic control techniques** for domestic heating, production of domestic hot water, setting and control of vehicle engines and electric motors, control of energy-intensive production processes;

- improved **production processes;**

- greater use of **renewable energy** sources;

37.   Most of these savings could be achieved using technologies already available in the Community which could easily be used commercially on a wide scale. This would mean developing new industrial subsectors and especially setting up and organising the maintenance and servicing which are essential for ensuring that new plant and equipment are properly monitored. By 2000 this could lead, according to a study commissioned by the Commission, to the gross creation of 200 000 to 250 000 new jobs.

38.   In view of all these considerations, it is very difficult to quantify the precise effects on different fuels. The reductions in final consumption would mainly affect liquid and gaseous fuels, but also electricity, which would in turn affect the demand for solid and nuclear fuels through power station consumption. Overall, energy imports could be expected to decline by 70 million to 90 million toe, which would reduce the Community's dependence on imported energy to 42%.

39.   It should also be pointed out that if lower prices on the energy market were to lead to a more "laissez faire" policy, the Community's energy consumption could substantially increase. Unless current energy savings policies are continued (with building-sector regulations, support for the development and marketing of energy savings technologies and so on), even the savings suggested in the reference projection would not be possible. This would have major repercussions on the level of primary energy demand and on imports, and consequently on prices and the level of economic activity.

*

*   *

CONCLUSIONS

1. "Energy 2000" outlines a plausible socio-economic scenario for the
    European Community and analyses the energy market consequences. It
    goes on to consider the sensitivities to changes in the key
    assumptions. Changes not studied are those which might be engendered
    by strictly political decisions or by a future crisis situation.
    Even so, the overall results of the study cover a sufficiently wide
    range of possible outcomes to identify the main elements likely to
    determine the Community's energy market in 2000.

    These elements were taken into account in the preparation of the
    Community's new energy objectives for 1995[1]. The most important
    findings are given below.

2. The degree of success likely to be achieved by the Community
    countries in making **more efficient use of energy** is by far the most
    important factor in influencing future energy requirements. The
    impact on the energy market of today's uncertainties in this area is
    far greater than those stemming from faster or slower rates of
    economic growth in the long term.

3. The **protection of the environment** will be a major concern as the
    century draws to a close. Accordingly, the Commission's departments
    are already urging the adoption of a Directive to limit the emission
    of pollutants into the air from large combustion plants. Application
    of the standards proposed should not affect the overall level of
    energy consumption but could give rise to significant trade-offs
    between energy products. And the financial impact of these measures
    both in the energy producing sectors and among consumers will
    certainly have some repercussions for energy prices.

4. **Stable or moderately declining energy prices** will not necessarily
    militate against the restructuring of the energy market in the long
    term. They may have a stimulating effect on GDP and, through
    appropriate policies, they could be a spur to investment. The more
    rapid turnover of the capital stock, bringing with it more
    energy-efficient processes could offset much of the increase in
    consumption that could be directly triggered by the effect of lower
    prices on demand.

    Sharply fluctuating energy prices, however, would disturb the smooth
    adjustment process in the Community energy economy, creating
    discontinuity in the trends observed in recent years. And sustained
    low prices would carry the needs of future price hikes in the
    absence of countervailing measures to keep the process of
    restructuring under way.

---

[1] Document COM(85)245.

5. A key factor in the restructuring of final energy demand will be the **penetration of electricity.**   This should be the fastest growing energy carrier from now until the end of the century, despite a ratio of electricity demand to economic growth of less than 1.

    Additional costs resulting from stricter emission standards for power stations should have little effect on final electricity demand, but could strongly affect the structure of energy consumption for heat and electricity generation and constrain the planned return to coal in this sector.

    The balance between nuclear energy and solid fuels for electricity generation in the Community could be modified as a result; but even if this happens the complementary roles of coal and nuclear energy in power stations are likely to be confirmed.

6. In the residential and tertiary sector, the **substitution of natural gas and electricity for oil products** will make an important contribution to change. The increased penetration of natural gas in a sector subject to major seasonal fluctuations also implies increased supplies of gas to sectors such as industry whose offtakes are either more regular or more easily modulated.

7. Successful maintenance of structural change in the energy economy should help to stabilize **consumption of oil products.** In that case these will be increasingly concentrated on uses where there are no real substitutes. The shares of light and medium products will therefore increase in the medium term at the expense of heavy products, a movement which would warrant continued rationalisation efforts in the refining sector.

8. Continued efforts to diversify supplies will mean replacing oil, most of it imported, by **solid fuels or natural gas, a growing share of which will also have to be imported in the future.** Any excessive dependence on a single origin could pose in the longer term problems similar at least in nature to those created by the dependence on oil in the early 1970s, but dependence on imported gas for the Community's total energy requirements will not be on a similar scale to that of oil.

9. Any development, however marginal, of the **Community's own energy production** under satisfactory economic conditions will help to reduce the potential vulnerability associated with import dependence. Hence the importance of sustaining the pace of nuclear programmes, developing new and renewable energy sources, maintaining a reasonable solid-fuel extraction level and continuing hydrocarbon prospection and exploration, which could eventually mean new fields coming on stream.

10. **The world energy market should** produce sufficient volumes of energy to cover the Community countries' import requirements. But the future oil market will be strongly influenced by consumption trends

in the other industrialized countries, especially the United States,
and in the developing countries. It will therefore be imperative for
the industrial partners in the Community to continue their every
effort to reduce their dependence on imported oil and for the third
world countries, especially those importing oil, to be able to
benefit from greater diversification of energy sources and a more
rational use of energy. Sustained low oil prices could in time
create the conditions under which serious pressures on world oil
supplies emerge in the absence of continuing attention the process
of energy diversification and greater efficiency of energy use
world-wide.

*   *

*